Junkers Airplanes of WWI

Volume 1: J.1–J.4

A Centennial Perspective on Great War Airplanes

Colin A. Owers

AF880042

Great War Aviation Centennial Series #30

This book is dedicated to the late Barbara Huston whose love of WWI aviation history was only exceeded by her love for her husband, Colin Huston.

Acknowledgements

This work would be much poorer were it not for the generous help provided over many years from friends, old and new, some sadly no longer with us. Any errors are mine alone.

Thanks to Gregory Alegi; Richard Alexander, Lennart Andersson; Garry Barling; Klas Bask; Carl Bobrow, Amy Brennan; the late Jack Bruce; Moshe Bu, Jean-Luc Claessens; Christophe Cony; Matt Dravitzki; Juanita Franzi; Roberto Gentilli; the late Peter M Grosz; Jack and Nicol Herris; Colin and Barbara Huston; Sir Peter Jackson; Roland Jahne; Stuart Leslie; Piotr Mrozowski; David Ostrowsky; Gennady Petrov; Charles Schaedel; Bruno Schmäling; the late Ian Stair; Gennady Sloutskiy; Chuck Thomas, William Bill Toohey; John Weatherseed, The Vintage Aviation Team; Aaron Weaver; Greg VanWyngarden; Reinhard Zankl; the staff of the RAF Museum; the British and US National Archives; the Deutsches Technikmuseum, Berlin; the Technikmuseum "Hugo Junkers", Dessau; and the Aviation Heritage Trust (AHT). Especial thanks to my wife, Julie, who accompanied me to the various archives and copied many of the documents used to develop this history. The colour scheme drawings by Piotr Mrozowski are based on his research.

Cover painting by Steve Anderson. Please see Steve's website at: **www.anderson-art.com**

Color aircraft profiles © Juanita Franzi. Scale drawings by Marty Digmayer.

For our aviation books in print and electronic format, please see our website at: **www.aeronautbooks.com**. You may contact the publisher at **jherris@me.com**.

ISBN: 978-1-935881-65-0

Design and layout: Jack Herris
Cover design: Aaron Weaver
Cover painting: Steve Anderson
Color profiles: Juanita Franzi
Scale drawings: Marty Digmayer
Digital photo editing: Aaron Weaver & Jack Herris

Aeronaut Books

Books for Enthusiasts by Enthusiasts
www.aeronautbooks.com

Table of Contents

Below & Facing Page: The Junkers technical museum at Dessau was compiled by Junkers after the end of WWI and gave a rich display of original Junkers material, wind tunnel models, and full-size machines covering the whole spectrum of Junkers aeronautical endeavours. In this photograph can be seen the remains of a J 2 monoplane suspended on the left and a J 4 fuselage on the right-hand side. The large displays included original models and parts showing the metal-working techniques pioneered by the Junkers team.

Foreword

Germany made more and better use of metal as a construction material in its aircraft than the Allies during WWI. The *Report on German Aircraft Construction* released in January 1918 by the British Ministry of Munitions of War – Department of Aircraft Production, while describing the multi-ply fuselage construction used by the Germans neglects to cover the steel tube fuselages developed by A.E.G. and Fokker.[1] The lack of any mention of Junkers aircraft in the report is understandable as none had been examined at that time.

Claude Dornier was working at the Zeppelin-Werft-Lindau (ZWF) facility using metal spars and developed the Rs series of large flying boats culminating in the Rs IV of late 1918. He also used stressed skin on his experimental 1917 CL.I biplane as a test for the Rs IV fuselage. At the end of the war Dornier had produced the D.I biplane fighter with metal cantilevered wings. Like all Dornier's efforts the structure had fabric covering on wing and tail surfaces over a metal structure. The US was so impressed with the D.I that they ordered two post-war, one for the Army and one for the Navy.

Professor Hugo Junkers followed a different path and was to develop a new branch of aircraft manufacturing – the series production of the all metal aeroplane. Apart from the problems he had

1 Copy in TNA MUN4/5010.

Above: Dornier Rs I flying boat. Construction of this gigantic flying boat was commenced in January 1915. The hull longerons and wing spars were manufactured from high-tensile steel. Ribs were metal and the bottom of the hull was covered with alloy sheet. The three 240-hp Maybach Mb IV engines were buried in the hull and drove the pusher propellers through shafts and bevel gear transmission. The flying boat was wrecked before flight trials could commence. Dornier's efforts were more in the line of an essay in production techniques rather than a desire to produce a production version. His results were encouraging and Count Zeppelin continued to support his efforts.

Above: The Zeppelin-Dornier Rs IV monoplane flying boat was manufactured between January and October 1918, when it was launched. The hull had sponsons, a feature that became a speciality of Dornier flying boats. The four 245-hp Maybach Mb IVa engines were mounted in power eggs between the hull and the high slung fuselage, one tractor and one in pusher configuration. The fuselage was constructed of metal with steel longitudinals, alloy frames and was metal covered. Only *Marine Nummer* 8801 was completed before the Armistice. It was dismantled in 1919.

Above & Below: The Zeppelin-Dornier CS.I was an experimental floatplane built during 1918 as evidenced by the style of national marking. It was constructed completely of metal except for the fabric-covered wing and tail surfaces. The machine had the V8-eight-cylinder Benz Bz IIIb engine and a normal Hussmann gun ring for the observer, a German copy of the Scarff mounting. No production was undertaken, only *Marine Nummer* 8502 being built.

with designing new manufacturing techniques to allow for the use of metal as the structure of the aircraft, he also had to convince officials in the German military of the worth of his ideas and then to fight to retain manufacturing at his Dessau facility. When it is considered that, due to his liberal ideals, Junkers was thought of as a radical by the conservative establishment, it is a wonder that he achieved so much during WWI.

With the help of original documents, German, British, and USA archives, the author has tried to put together a complete coverage of the aircraft built and those proposed during the period of WWI. As with all WWI aircraft records vary, particularly with regard to dimensions. This is especially so with the Junkers J 8/J 10 monoplanes. The specification tables record the dimensions from original sources where possible, compared with modern sources so that the reader

Above: Zeppelin-Dornier D.I. This single-seat fighter was Dornier's experiment into cantilever wings. About two thirds of the wings were covered with aluminium over alloy ribs and spars. The demarcation line of the fabric and metal on the wing is clearly shown in the photograph.

Above: The fuselage was an all-metal construction featuring stressed skin. The single leg undercarriage and underslung jettisonable fuel tank were not to be common until much later. The machine suffered a catastrophe when *Hauptmann* Reinhardt, commander of *Jagdgeschwader Nr.*1, suffered wing failure on D.1751/18 during a flight on 3 July 1918, at the Second Fighter Competition at Aldershof. The top wing tore off the machine, Reinhardt suffering fatal injuries in the subsequent crash.

Above: The US Navy's Dornier D.I in the Naval Aircraft Factory, Note the fabric covered surfaces. This machine was acquired from a Swiss agent. In one document, it states that the machine was a seaplane. It has not been possible to determine if the two requisitions attributed to this aircraft covered the Army and Navy's machines. Given the Bureau No. A-6058 by the Navy it was enroute in late August 1921 and shipped to the Naval Aircraft Factory on 31st of that month. No reports of what was done with the machine have been discovered to date and it was at the NAF when it was surveyed (NAF 182-23) and written off on 9 November 1923.

Below: Bearing the US Army serial A.S. 68546 and the McCook Field P-Number P-241, this Dornier D.I was placed in the McCook Field museum on 14 May 1923. Its 200-hp this BMW engine (Serial No. 94640) had been loaned to the ED Stimson Co of Detroit on 16 May 1922. The engine was returned on 22 March 1926. A survey was undertaken on 8 September 1926, but it does not state whether this is the engine or airframe. For more on the Dornier construction see *Dornier Metallflugzeuge 1914–1919*, by Michael Schmeelke; Aeronaut is planning to publish an English edition.

Above: A Junkers J 4 (Military J.I) in a field with what appears to be a J 9 (Military D.I) fighter to the left rear and a J 10 (Military CL.I) two-seat fighter on the right. Junkers aircraft were completely metal with the exception of a few items like tail skids and engine bearers.

can see where the latter have originated from and to be able to determine what source they consider more correct.

Another problem with identifying Junkers aircraft is the fact that the factory used J numbers from the J 1. When the Army introduced the armoured infantry cooperation aircraft as a new type in 1917, the J-type designation came about due to the term "I" for *Infanterie* that in old German looks like a "J", and this was adopted such that it is always written as a J today. The official Army designation for the J 4 was the J.I, and this has caused confusion in some publications.

The J 4 (Military J.I) had proved itself on the battlefield but it was a dead-end. It was to be in the post-war unrest on Germany's borders that Junkers military aircraft would prove themselves. The techniques used in the J 9 (Military D.I) and the J 10 (Military CL.I) were to be carried on into the Junkers civil airliners that led to the successful Junkers Ju 52 tri-motor transport of the 1930s. From his earliest days Professor Junkers perused the concept of a large aircraft with the passengers accommodated within the wings. He believed that only with a large size could the combination of metal construction and the suppression the drag of ordinary fuselages be successfully attained.

Facing Page: After the Nazis took over the Junkers concern they kept the name due to its international reputation for tough, reliable aircraft. This book, *Die Junkers Lehrschau*, was produced just before World War II and provides a history of Junkers products and shows the Junkers technical school museum. Appropriate pages from this book are reproduced throughout these two volumes. The book cover shows the company logo *Der fliegende Mensch* (The Flying Man) designed in 1925 by F. P. Drömmer.

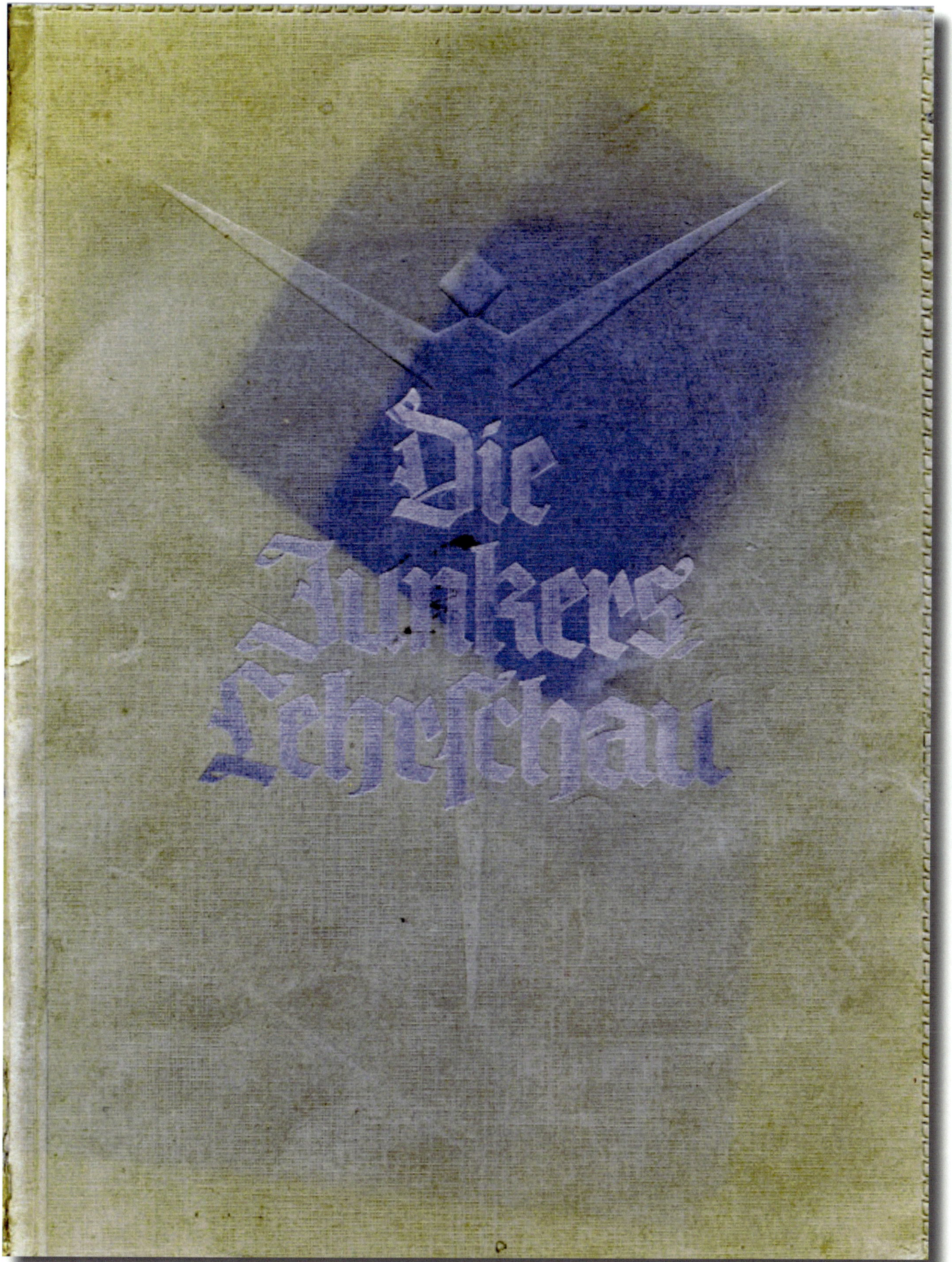
Die
Junkers
Lehrschau

1. Hugo Junkers - Aeronautical Pioneer

In 1923 Professor Junkers was to give a paper to the Royal Aeronautical Society in Great Britain, but he was unable to attend due to illness. *Ing* Hermann Mierzinsky, a consultant engineer at the Junkers Dessau factory and the professor's confidential secretary, attended in his stead. W.J. Stern of the Air Ministry translated and read the paper for him. In describing the method that led to his all metal construction of aeroplanes Professor Junkers stressed the importance of co-operation between the research and production sides of the endeavour.[1] The problems of producing Junkers all-metal aircraft during the war no doubt led him to this conclusion. The paper included photographs and plans of the early wartime Junkers products as well as future designs where the passengers were wholly enclosed within the thick wings to make the most of streamlining.[2]

Hugo Junkers was born on 3 February 1859. The Junkers family can trace their origins back to the 16th Century. Earliest records show that they were farm workers near Düsseldorf. In 1818 a textile company was established by Johann Peter Junkers, Hugo Junkers' grandfather. The textile company was carried on by Hugo's father and a brother until the 20th Century. Hugo graduated from the Berlin University in 1888 and went to work with the Continental Gas Co of Dessau. He left to form his own company in 1897 to manufacture gas water heaters. He also lectured at Aachen University and was endowed with a professorship in thermodynamics. Married in 1898 at the age of 39 to Therese Bennhold, they had twelve children, the last being born in 1920. Junkers was described as energetic, brilliant with a tremendous capacity for work. He started the Junkers Motor Works and developed a double piston stationary diesel engine. Junkers devised and patented many different inventions in relation to gas heaters for industrial and domestic work, gas stoves, oil and petrol engines, the profits from which enabled him to pursue his interest in aerodynamics at the age of 55.

Junkers became interested in aviation in 1909/1910 in association with Hans Reissner a mathematics professor at the *Technische Hochschule Aachen* (Technical High School at Aachen). Reissner invented the first German variable pitch airscrew. In 1912 the wings, radiator, elevator of a canard type monoplane with single surface aluminium wings to the specifications of Reissner, were constructed at the Junkers water heater works

Above: Professor Hugo Junkers.

in Dessau. The aircraft was of canard configuration and named the *Ente* (duck). The first Reissner machine made its maiden flight on 23 may 1912, but was damaged later in an accident. It is unknown if it was repaired or if the second Reissner machine was an entirely new design. This aircraft first flew in September 1912 from Berlin-Johannisthal. It crashed on 27 January 1913 but was repaired afterwards. Between then and 1915 Junkers undertook his own experiments with his ideas on aerodynamics and eventually for constructing a metal aircraft. He constructed a wind tunnel at the Technical High School at Aachen in 1911 wherein he conducted his experiments on model sections as he developed his understanding of aerodynamics. On leaving the School in 1912 he built a wind tunnel at his Dessau factory as he wanted to carry on his aerodynamic experiments without other commitments. Junkers worked to eliminate the parasite drag that was inherent in the biplane. On 1 February 1912, Junkers had been granted German patent No. 253788 covering his idea of enclosing the motor, passengers and all structural members within the wing

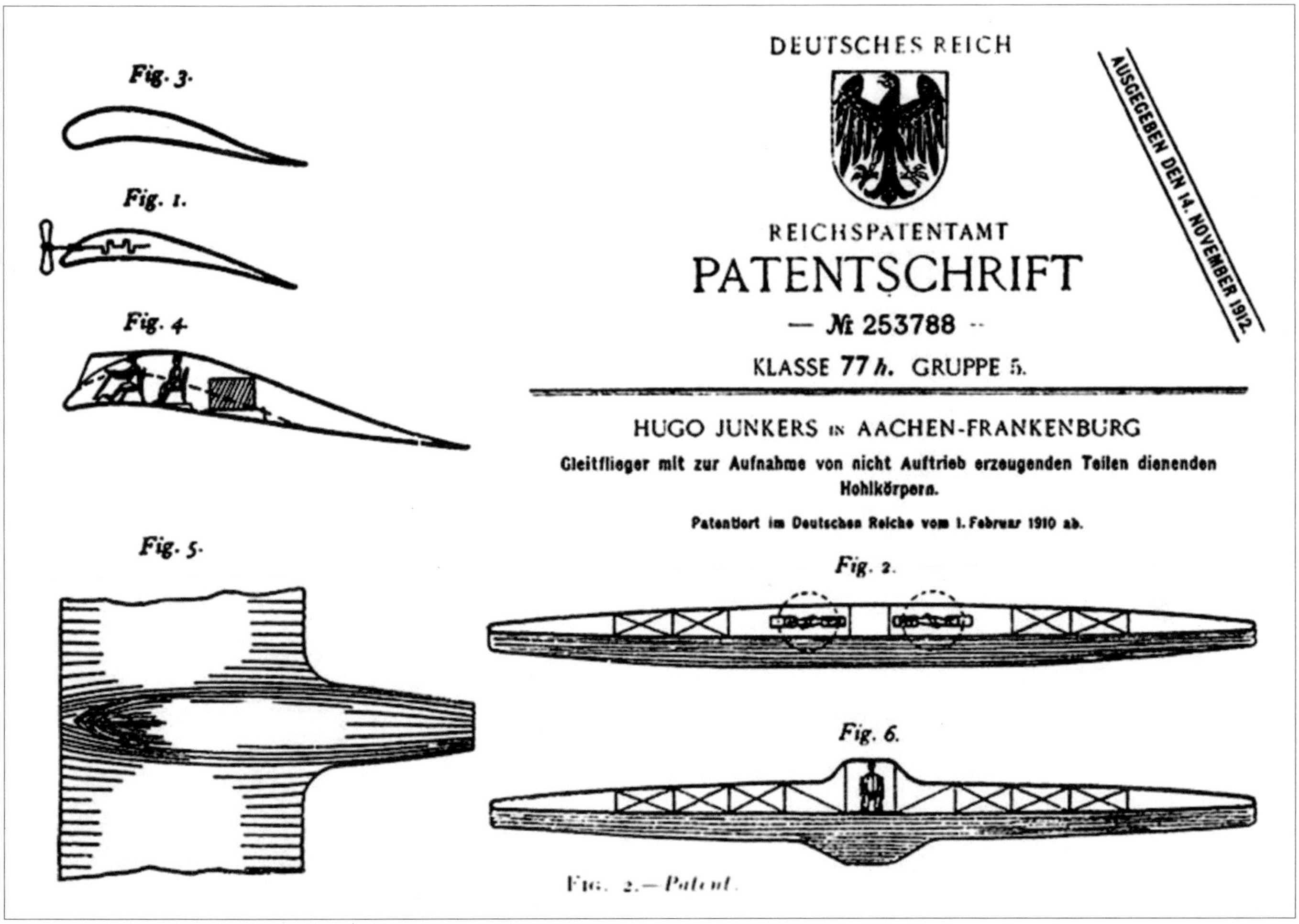

Above: Sketches from Junkers' 1910 patent for a flying wing.

Below: The Junkers wind tunnel at Aachen. (Junkers Album via R Zankl)

Above: The Junkers wind tunnel at Aachen.

Facing Page, Top: Details of the Junkers wind tunnel. The bottom two photographs show what appears to be radiators.
Facing Page, Bottom: The wind tunnel as it is today.

thereby eliminating all unnecessary drag. Naturally this required a thick wing and Junkers began the laboratory work necessary to prove his theories and to develop a cantilever wing. He finally settled on a family based on a 20% thick root section tapering in thickness with a constant mean camber.[3]

Junkers was a heating engineer and while shaping and bending sheet metal for ducting he came to the conclusion that the same techniques could be used to form metal into aircraft sections. Apparently, he was well on his way to building such an aircraft at his Dessau works when it was interrupted by the outbreak of war. Other sources state that he was too involved in many other projects to devote the time to aviation until the outbreak of the war. The Junkers plant was soon in full production of its products, especially of field stoves for use at the front.

Writing in his notebook in 1915 Junkers noted that to "build an aeroplane totally out of metal was a problem that seemed unsolvable.

> My friends wonder: What has gotten into Junkers that he keeps himself occupied with such unrealistic plans? Iron does not fly, an aircraft has to weigh light."[4]

In May 1915, Professor Junkers, with Dr Otto Mader and *Oberingenieur* Hans Strudel, approached the military with the proposal to build an all-metal aircraft. The reaction of the authorities in turning down the proposal is understandable. Here was a bath-stove maker without any airfield or pilots, who had not built an aeroplane, wanting to build a monoplane, and one of metal! The authorities were only interested in biplanes and considered metal unsuitable for a light aeroplane.

Junkers arranged for the military to inspect his

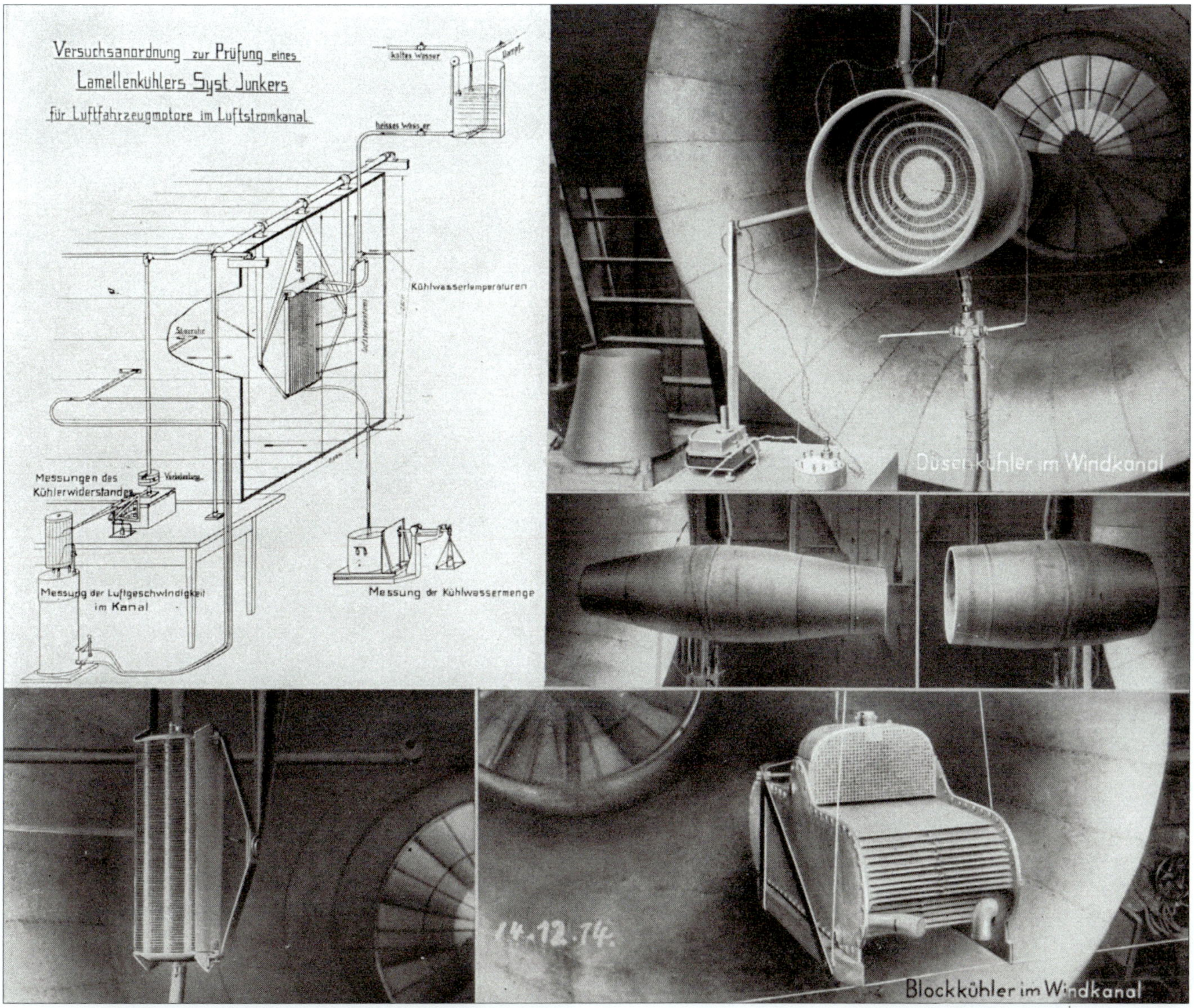
Versuchsanordnung zur Prüfung eines
Lamellenkühlers Syst Junkers
für Luftfahrzeugmotore im Luftstromkanal
kaltes Wasser
heisses Wasser
Kühlwassertemperaturen
Messungen des
Kühlerwiderstandes
Messung der Luftgeschwindigkeit
im Kanal
Messung der Kühlwassermenge
Düsenkühler im Windkanal
14.12.74.
Blockkühler im Windkanal

Above: Aerial view of the Junkers factories at Dessau. The numbers refer to the following: 1. Junkers original gas works. 2. Junkers & Co. 3. Junkers Flugzeugwerke A.G. 4. Junkers Motorenbau G.m.b.H. 5. Airfield. (AHT AL0032-14)

Above: The Junkers facility on 7 July 1919. See location 2 on previous photograph. (SDTB)

Above & Above Right: A model set up in the wind tunnel.

Below: Various model wings and designs were tested by Junkers in a systematic manner to develop his theories of aerodynamic cleanliness. Note the two trousered monoplane models on the top shelf with a model of the J 2 between them. (Junkers Album via R Zankl)

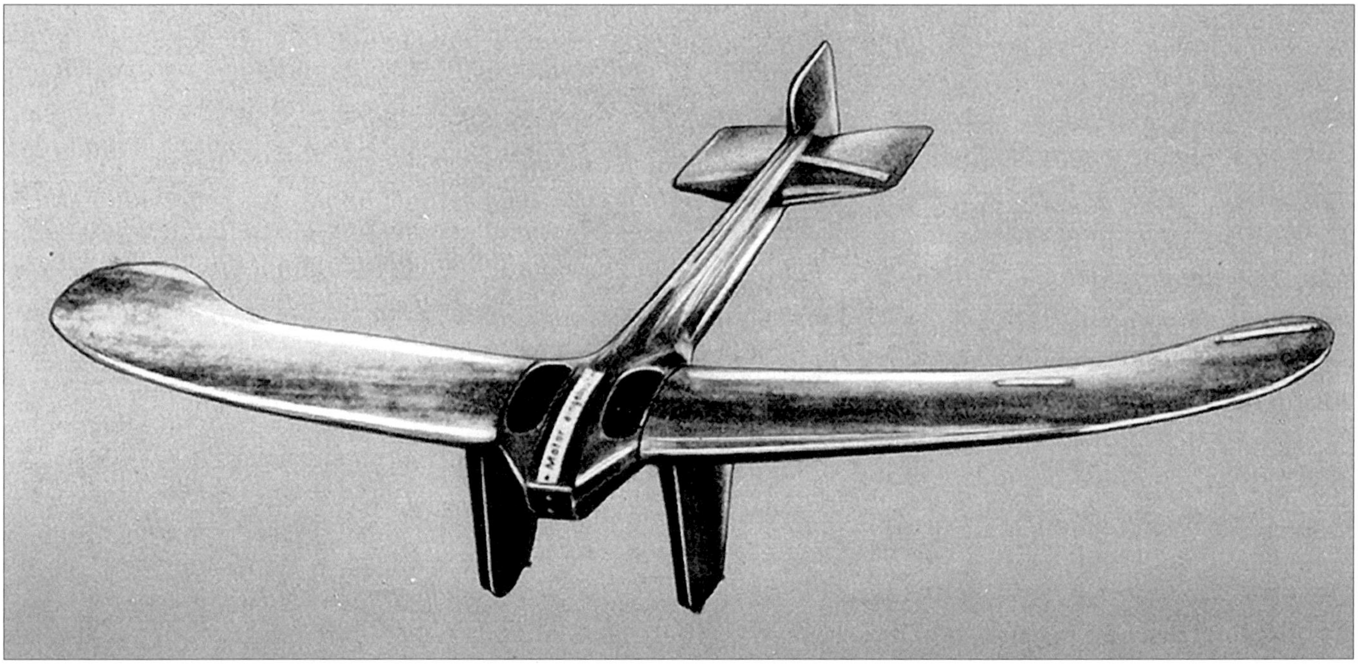

Above: A Junkers wind tunnel model from 1914 shows the advanced thinking of Professor Junkers. The best result from the hundreds of wind tunnel models at Aachen was a cantilever mid-wing monoplane with trousered landing gear. The trousered undercarriage legs were abandoned on the J 1 and a simple V-type undercarriage fitted.

Below: The aircraft workshop at the beginning of July 1915. A wing section is lying on the ground together with rib sections. (Junkers Album via R Zankl)

Above: The workshop at a later date.

Below: A Junkers experimental wing being tested under load. Otto Seifert, one of the supervisors, wrote that experimental wings were built first to test Junkers' theories. Duralumin was considered but could not achieve the necessary results. Working with small, thin sheets was extremely difficult and many problems had to be overcome.

Above: Second experimental metal wing of 9.2 m^2 surface.

Below: The third experimental sheet metal wing with a Fokker type load of 15 men certainly looked impressive. 6 October 1915.

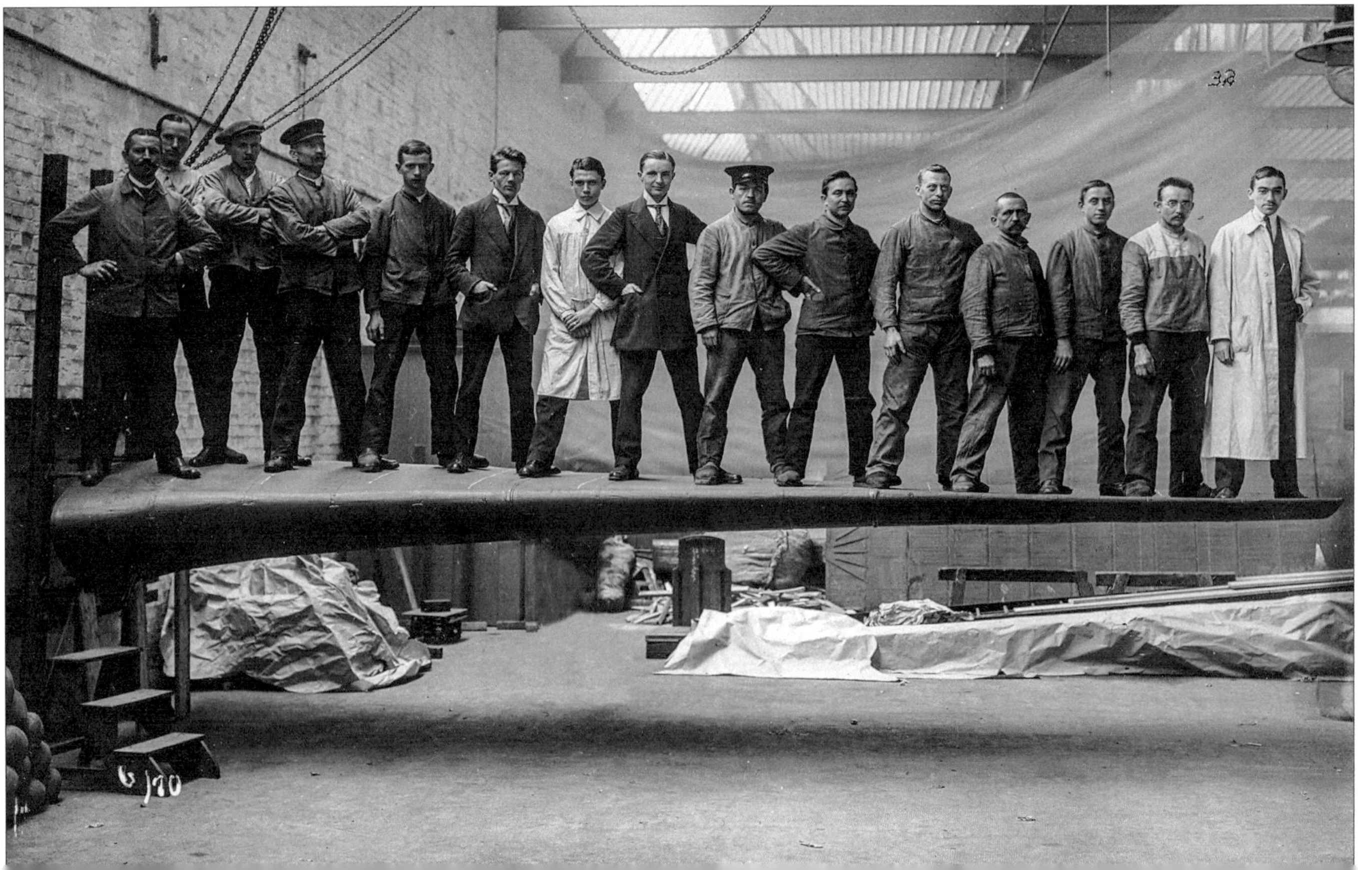

Above: The construction was such that two men could easily lift the wing.

Below & Following Page: The Junkers switched to lighter duralumin in the summer of 1916. This is one of several test wings constructed. The section could be carried by two men but could take a heavy load as shown with 100 and 200 kg/m^2. (Junkers Album via R Zankl)

Above & Below: Technical photographs showing loads of 100 kg and 200 kg per square metre. The wing was blanked off to hide the construction features until patent protection was granted. (Junkers Album via R Zankl)

Above: Technical photographs may be correct for reports but for the general press and public, a photograph of so many men standing on the wing was more impressive. (Junkers Album via R Zankl)

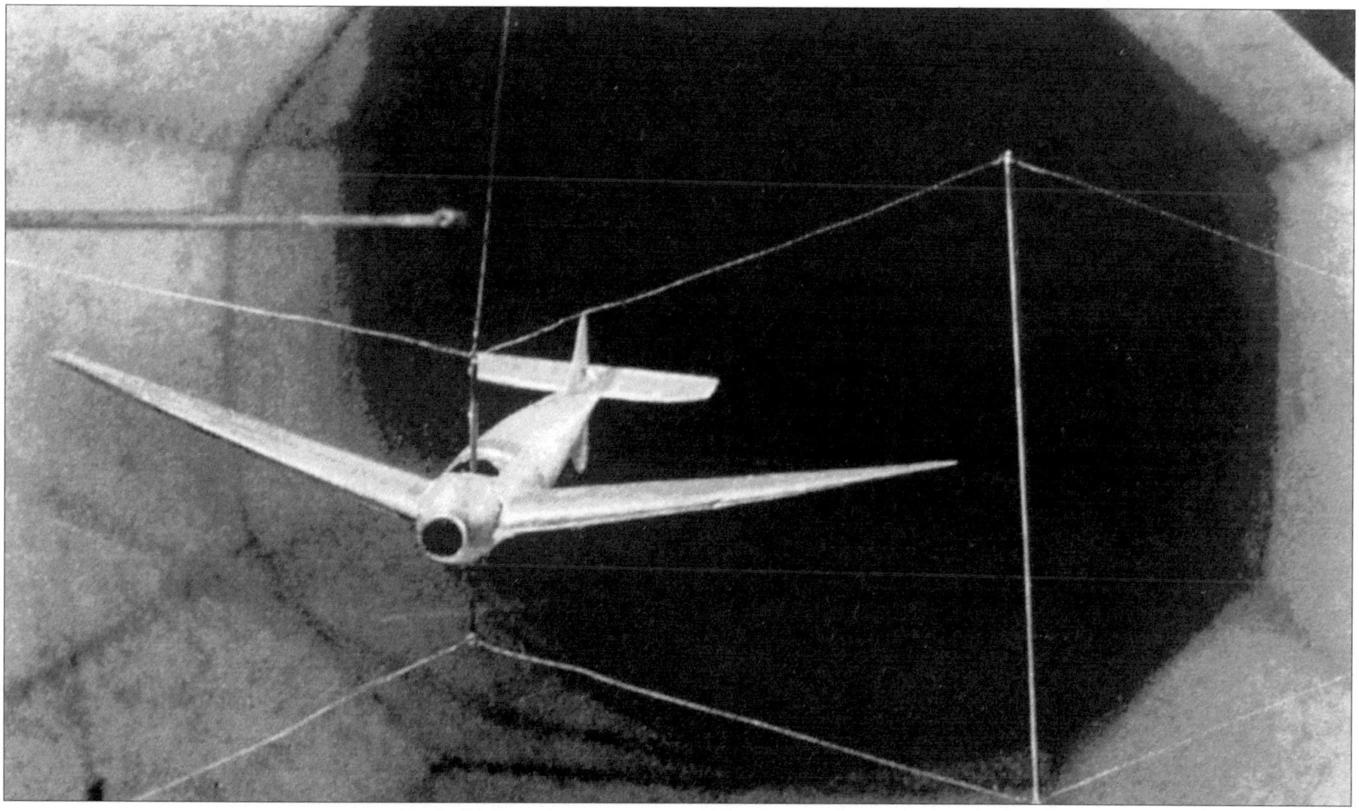

Above: A modern-looking wing-tunnel model from October 1916.

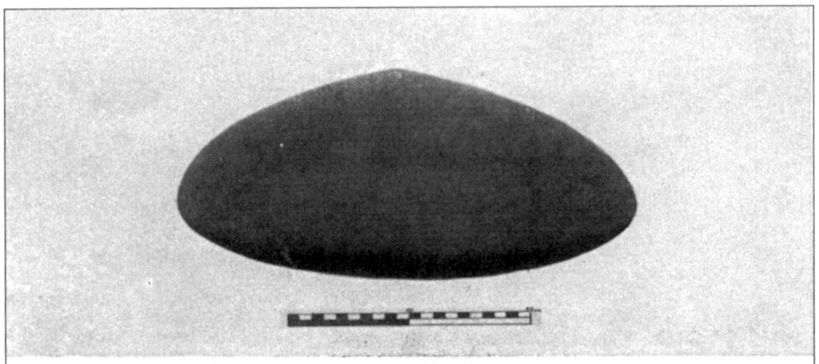

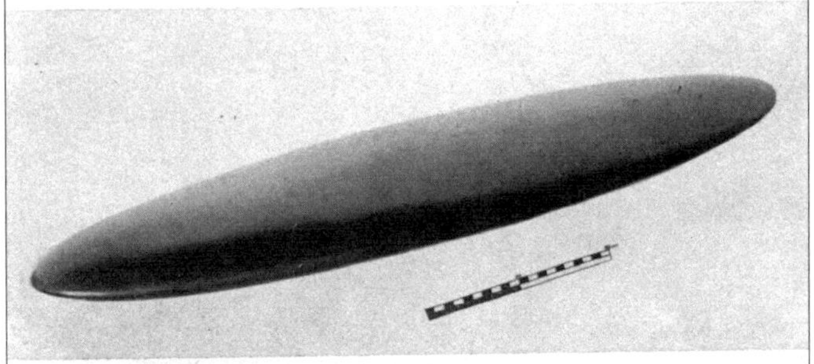

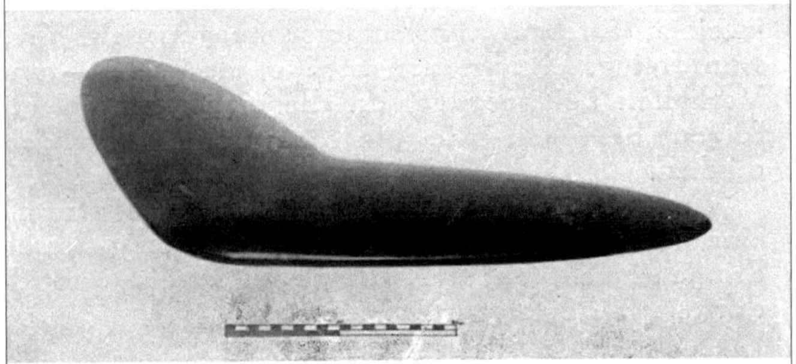

Above & Below: Some of the various shapes that were tried in the Junkers wind tunnel.

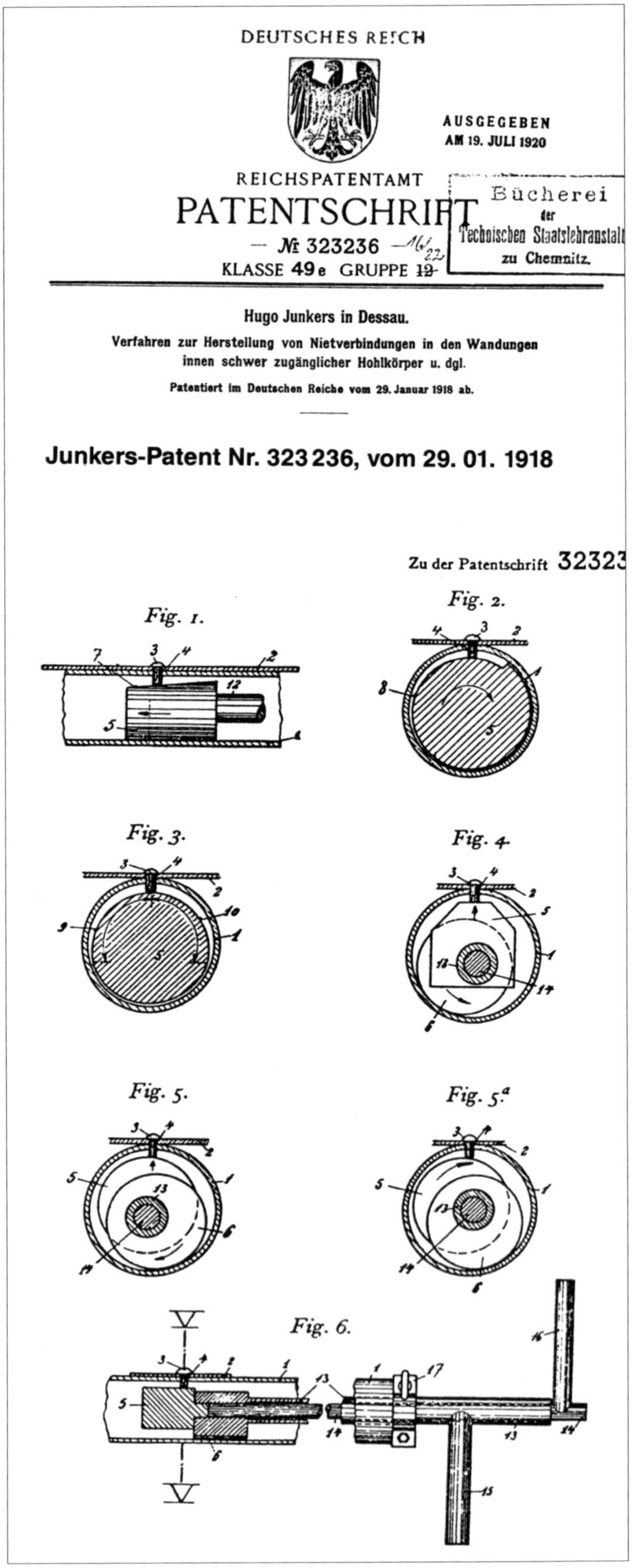

DEUTSCHES REICH

AUSGEGEBEN AM 19. JULI 1920

REICHSPATENTAMT

PATENTSCHRIFT

— № 323236 —

KLASSE 49e GRUPPE 19

Hugo Junkers in Dessau.

Verfahren zur Herstellung von Nietverbindungen in den Wandungen innen schwer zugänglicher Hohlkörper u. dgl.

Patentiert im Deutschen Reiche vom 29. Januar 1918 ab.

Junkers-Patent Nr. 323 236, vom 29. 01. 1918

Zu der Patentschrift 32323

Above: Junkers patent showing how they developed a method of riveting the metal skin to the metal pipe airframe.

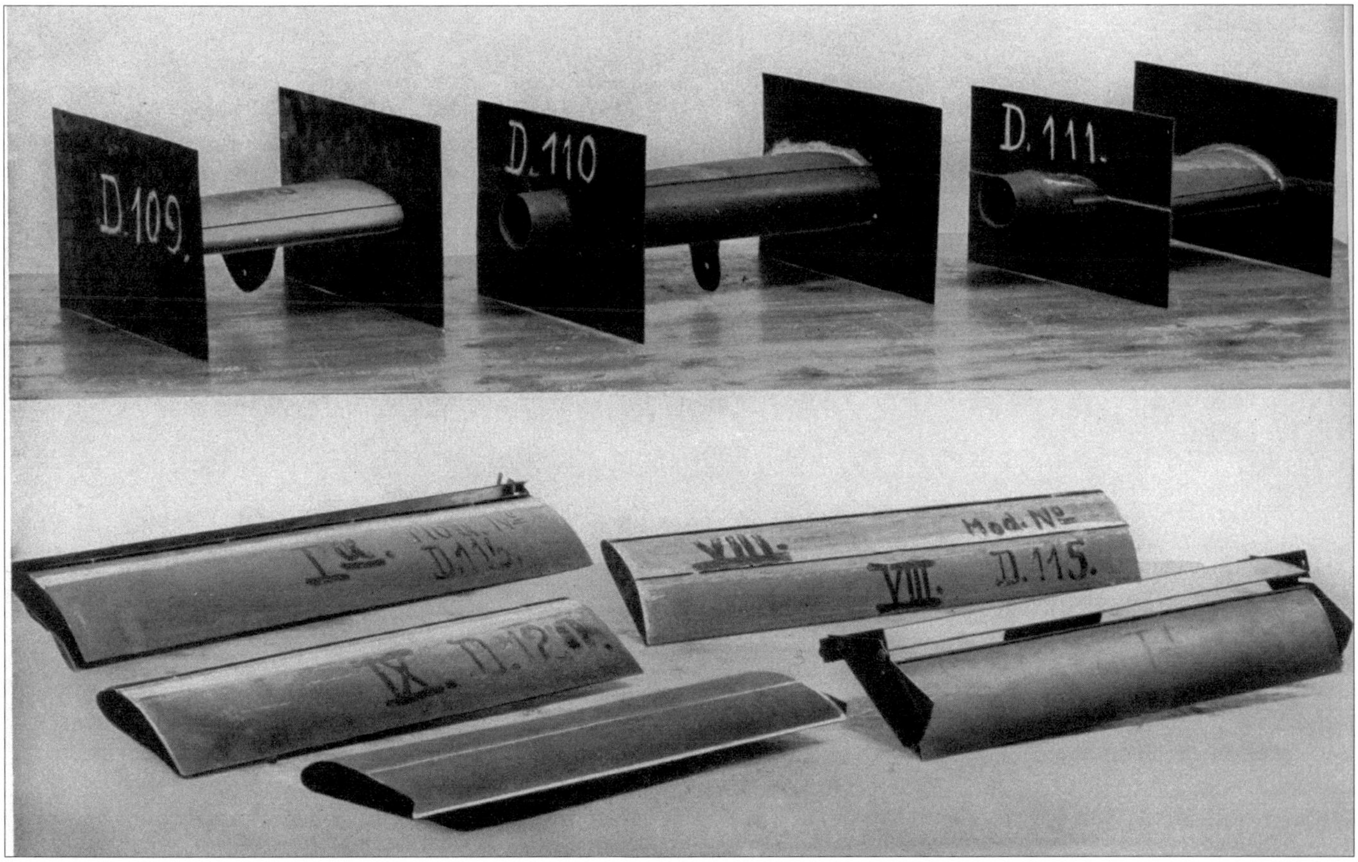

Above: Various aerofoil shapes developed by Junkers.

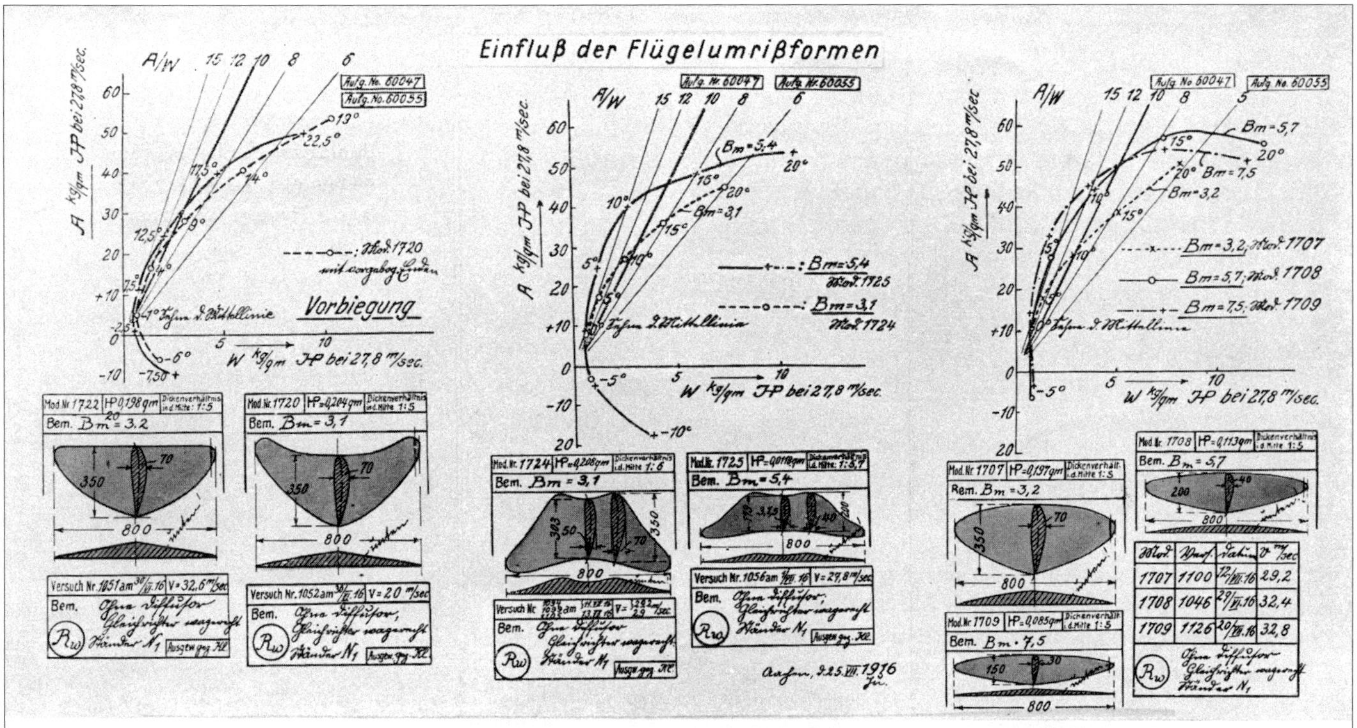

Beispiele von Versuchsergebnissen betreffend Einfluß der Grundrißform und des Seitenverhältnisses.

Above: The results of the testing in the Junkers wind tunnel.

Above: Sketch by Peter M. Grosz of proposed Marine DD all metal biplane that resembled that described here.

work. They were shown the metal working and spot welding machines, and the early tests of a metal wing. A visit to Aachen showed the wind tunnel and its results led to the military considering a provisional works programme for a metal aircraft, but it must be a biplane. They agreed to the monoplane as a test vehicle to prove the concept but a biplane must be developed and held in reserve.

Description of the Metal Aircraft of Prof. Junkers.[5]

The following unsigned document is also undated so where this proposed pusher two-seat monoplane falls in the history of Junkers aeroplanes is unknown. It is important as it gives an overview of the Junkers method of construction and what he considered the benefits of metal construction on aircraft design.

1.Differences Between Current Use Aircraft.

The aircraft differs in principle from the usual construction in two points:

a). Cantilever wing – no bracing wires or struts in the slipstream.

b). Construction material is only meta – no wood, no fabric not even for the wing covering.

Re a): Instead of the biplane layer construction held in the slipstream by struts and bracing wires, the wing itself is built to perform these functions. Therefore, thick wings at the root become necessary. By extensive trials in my own research facility I have discovered that there is no noticeable degradation in the thrust-lift-drag factor with this wing form. Also these thick wings provide greater stability.

The load bearing parts are separated as far as possible and this is taken up by the outer wing skin constructed from thin tin and fastened and arranged in a special way. In this way the wing, with minimum weight is able to withstand unusual forces as might be experienced during landing.

The removal of all struts and wires from the slipstream as well as the possibility to move other parts (for example fuel tanks) into the hollow wing reduces drag and thereby enables greater speed for the same power.

Re b): In this new form of utilisation metal has great advantages over the old wood-steel-fabric construction. It combines high strength in all directions (tension, compression and pressure) with elasticity and stretchability. It does not deform due to weather influences. It is capable of achieving any form and can be joined in any desirable way (welding, riveting, folding etc.). The dimension of the parts is not restricted to that of say wood. The larger the aircraft becomes the more advantageous metal construction becomes, just as the developments in ship building have shown.

2.Construction Details.

This proposal is based on a biplane of 45 sqm wing area, 14m span, and 150-hp engine with an efficient, effective pusher propeller. It shall have a two-man crew, machine gun in front, bombs in the middle, petrol tank in the wing above so that all the petrol can be used by gravity feed without pressurisation. All variable weights are placed near the centre of gravity to enable easy control at all times. The undercarriage is simple and should be fitted with special attachments to withstand hard landings.

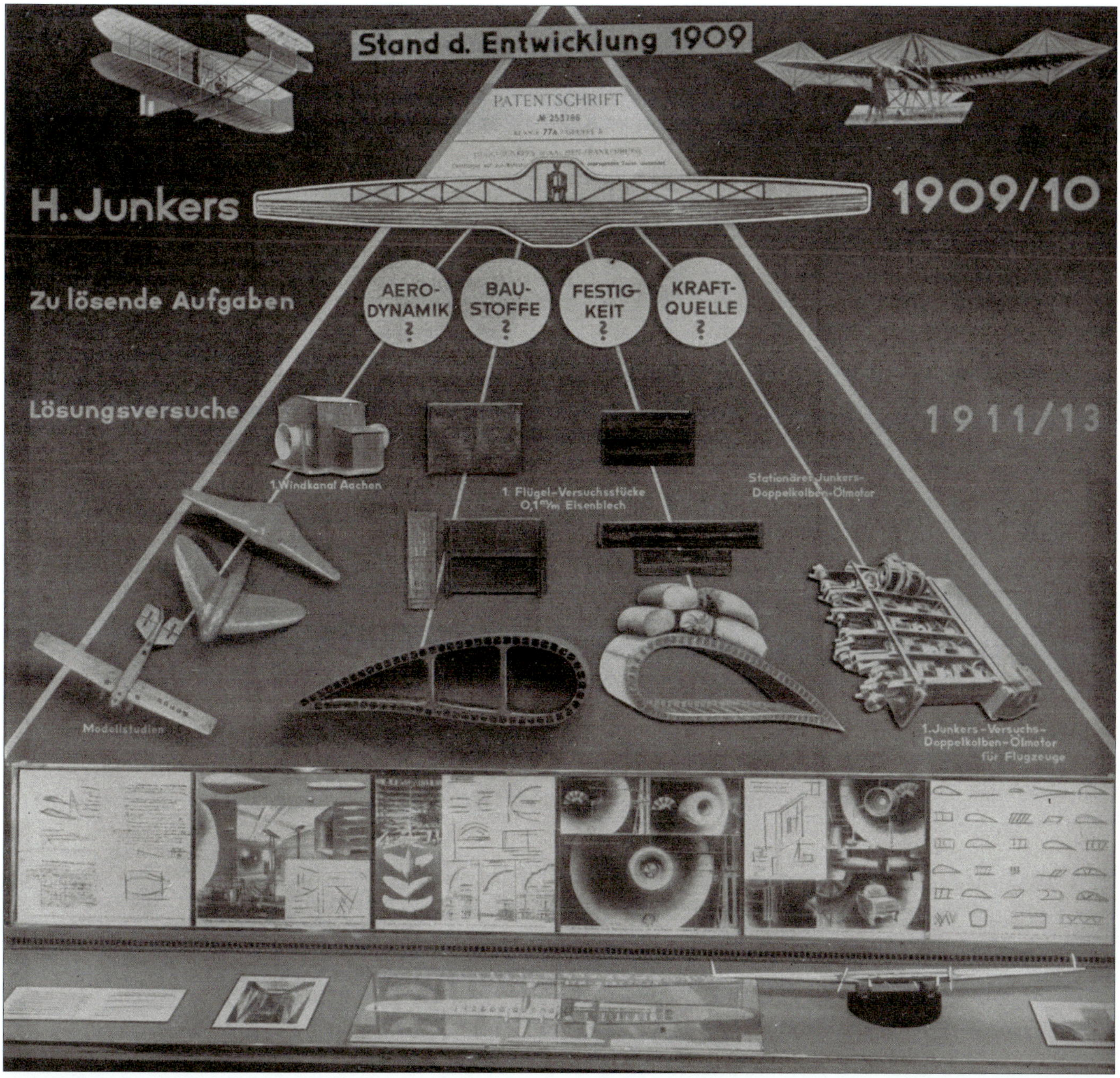

Above: This display at the Junkers museum between the world wars showed the development of Junkers ideas from the all flying wing to the J 1.

The disassembly of various parts for transport is achieved in the shortest time by undoing a few bolts leaving the machine unaffected as a whole.

3. Military Advantages Of The New Construction Method.

a). High Speed . Over 135 km/h due to removal of all bracing wires and struts, as well as the covering/ fairing of all parts where possible in loadbearing streamline form. All surfaces are smooth due to the metal construction.

b). Great strength. The material of the wings is so arranged that it acts as a single unit that when unusually stressed, such as in a hard landing, it can flex greatly before a single component breaks; also in the field it is unnecessary to adjust or re-tension the wing thus avoiding the danger of incorrect or negligent adjustments.

c).With battle damage the wings safety does not rely on a single strut or cable but even with larger areas of damage to the loadbearing skin the wing will not fail in normal flight.

Above & Below: Junkers structural method took considerable effort to develop. The tubes were not welded but connected by being riveted to steel brackets. As shown these could be quite complicated and contributed to the longer time a Junkers metal aircraft took to manufacture compared with a conventional wooden airframe. (SDTB)

Above & Below: These two photographs show the difference between the construction of the first Junkers all metal types and the adoption of the multi-tubular spar layout with corrugated skin.

d). Unrestricted view forward, sideways, to the rear and down for the observer and pilot.

e). Unrestricted field of fire, more than any other construction method, unrestricted forward by a propeller and sideways and to the rear by bracing. In larger types the machinegun could be replaced by a canon whose recoil is particularly well absorbed by the metal construction.

f). Housing a number of bombs at the centre of gravity of the machine without having an effect on control.

g). Petrol tank also at the cg but in the upper wing causing no drag and the petrol feeds by itself.

In June 1915 Junkers proposed a biplane to the *Reichsmarineamt (RMA)*. The surviving sketches of this project are similar to the description above.

Facing Page: The spar construction was carried through to the J 4 and following types. Here the corrugated covering has been removed to show the failure of the wing spar. (SDTB)

Right: Rolling the corrugations into a panel.

Endnotes Chapter 1

1. "Metal Aeroplanes – Résumé of Professor Junkers' Paper read before R.Ae.Soc." *Flight,* 11.01.1923.
2. *The Journal of the Royal Aeronautical Society,* Vol.27, P.406, 1923.
3. Sunderland, G. "Designing Fokker Aeroplanes", *Over the Front,* USA, Vol.30 No.3, 2015. P.215.
4. Information sheet. *Technikmuseum* "Hugo Junkers", Dessau. July 2017.
5. Original document via B. Schmäling.

Below: The forward fuselage section of a CL.I shows how Junkers overcame the problems with the wing to fuselage join by building the wing integral with the fuselage. Note the single machine-gun mount. (SDTB)

2. Junkers J 1 – The World's First All-Metal Aeroplane

In late 1914 or early 1915 Junkers again returned to the subject of aviation despite the demands of the war.[1] Work began at the Dessau laboratory where Junkers had established his *Forschungsanstalt* (research institute). This was a self-funded project at this time. Assisted by engineers Otto Reuter and Otto Mader, Junkers designed an experimental two-seat low-wing monoplane with a thick wing, the J 1. Junkers had tested some 3,000 to 4,000 different models in the wind tunnel in order to find the most favourable profile. The Junkers team also built and refined a number of experimental metal wings. As noted above, the *Kriegsministerium* (War Ministry) was approached in early 1915 to support such an aircraft and were so impressed with the work at Dessau that a contract was issued on 30 June 1915, for the experimental aircraft.

Construction of the actual aircraft started in September. It was a purely experimental machine to test Junkers concepts. It was built within four months and was entirely of steel, its unusual angular appearance earning it the unofficial title *Blech Esel* (Tin Donkey). According to some sources it had been intended to cover the steel tube structure with duralumin but wartime shortages forced the use of thin steel plate spot welded to the frame. Duralumin alloy had been developed by the German metallurgist Alfred Wilm and had been available from 1909. It was used in rigid airship frames. The process to produce the material was sensitive and it was prone to flaking and powdering in sheet form. Once the channelling of the duraluminium sheets was invented it became the material of choice for Junkers aircraft. The material available for the J 1 was small sheets of steel 0.2 to 0.5 mm in thickness.[2] Junkers wrote later that "sheet iron 0.1 mm thick used for magnetic purposes in plates of a very disadvantageous small size and quite unsatisfactory properties of resistance and elasticity" was all that was available.[3] A few important items were constructed from steel tubes. To make the plate rigid a second layer of corrugated sheet was welded to the inner side.

The J 1 was a mid-wing monoplane with the engine and cockpit section comprised of welded steel tubes. The box-like rear fuselage structure was composed of rings of welded sheet metal panels 30 to 50 cm in length on built up bulkheads of welded steel strip I-sections and angle sections with some steel tubing.

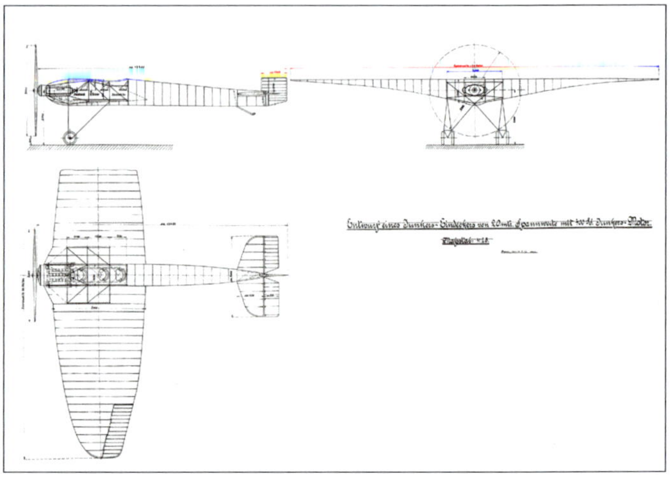

Above: Junkers drawings of proposed three seat monoplane. 10 Aug 1916 had the wing highly integrated with the fuselage. The wing planform was elliptical to minimize induced drag.

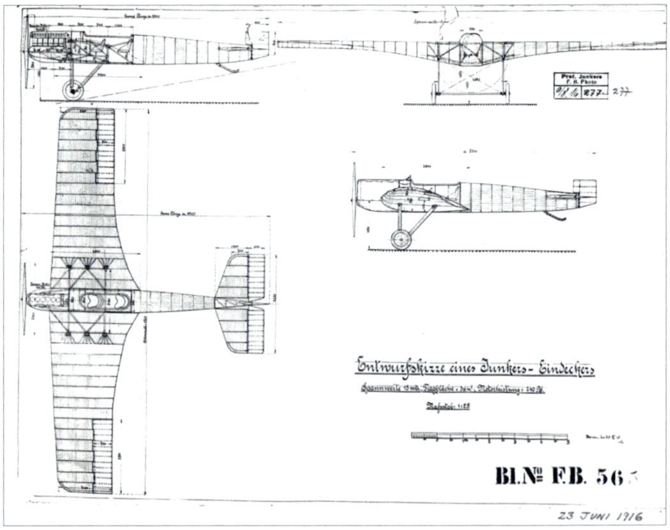

Above: Another Junkers proposal on the way to the J 1. The rear fuselage and tail shape have been determined but the wings required refining. Note the provision for two seats. 23 June 1916.

The fixed horizontal tail surfaces were cantilevered with plain elevators. The balanced rudder was mounted without a fin.

The landing gear was a simple welded steel tube construction with rubber shock cords, the

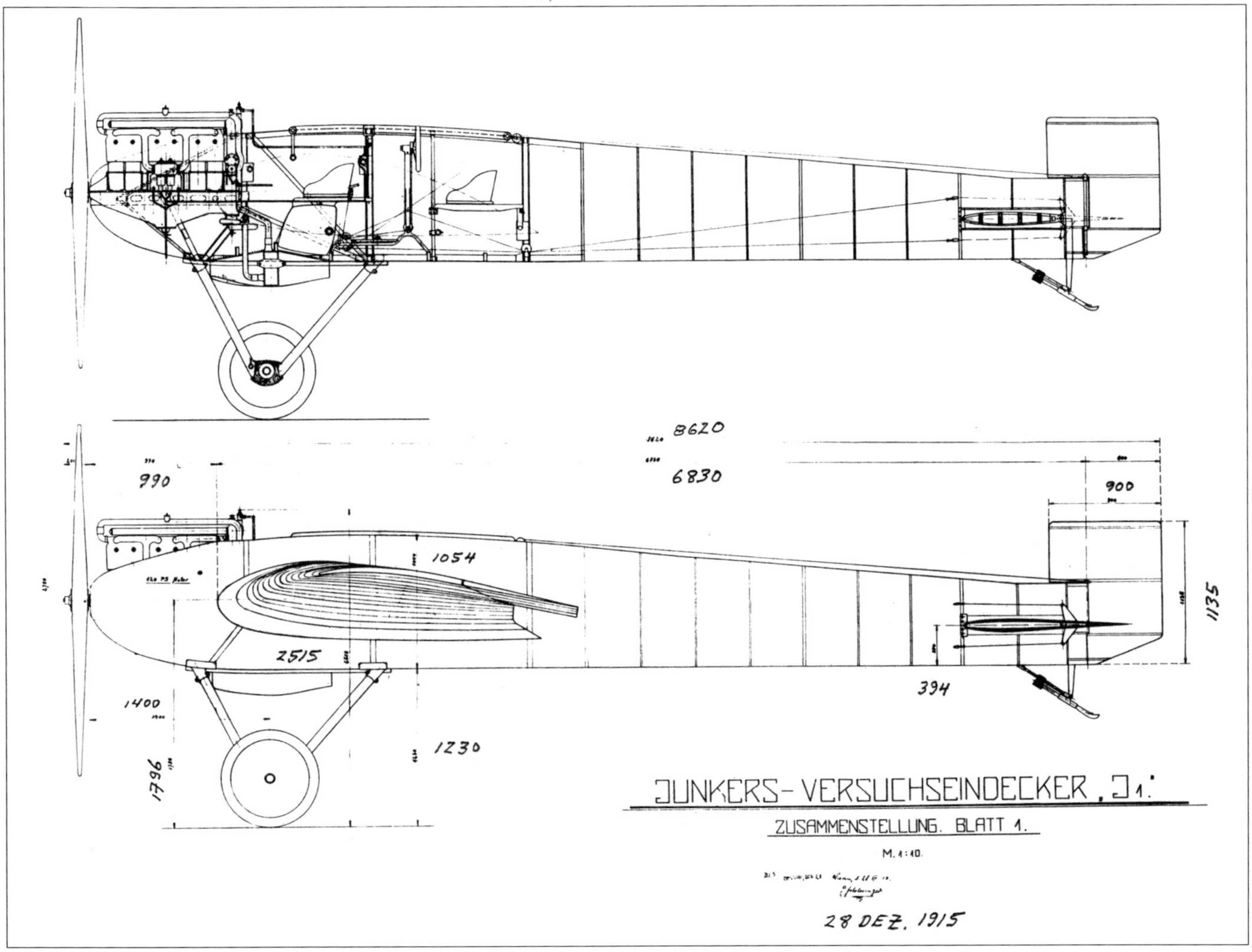

Above, Below, & Following Page: The J 1 as built.

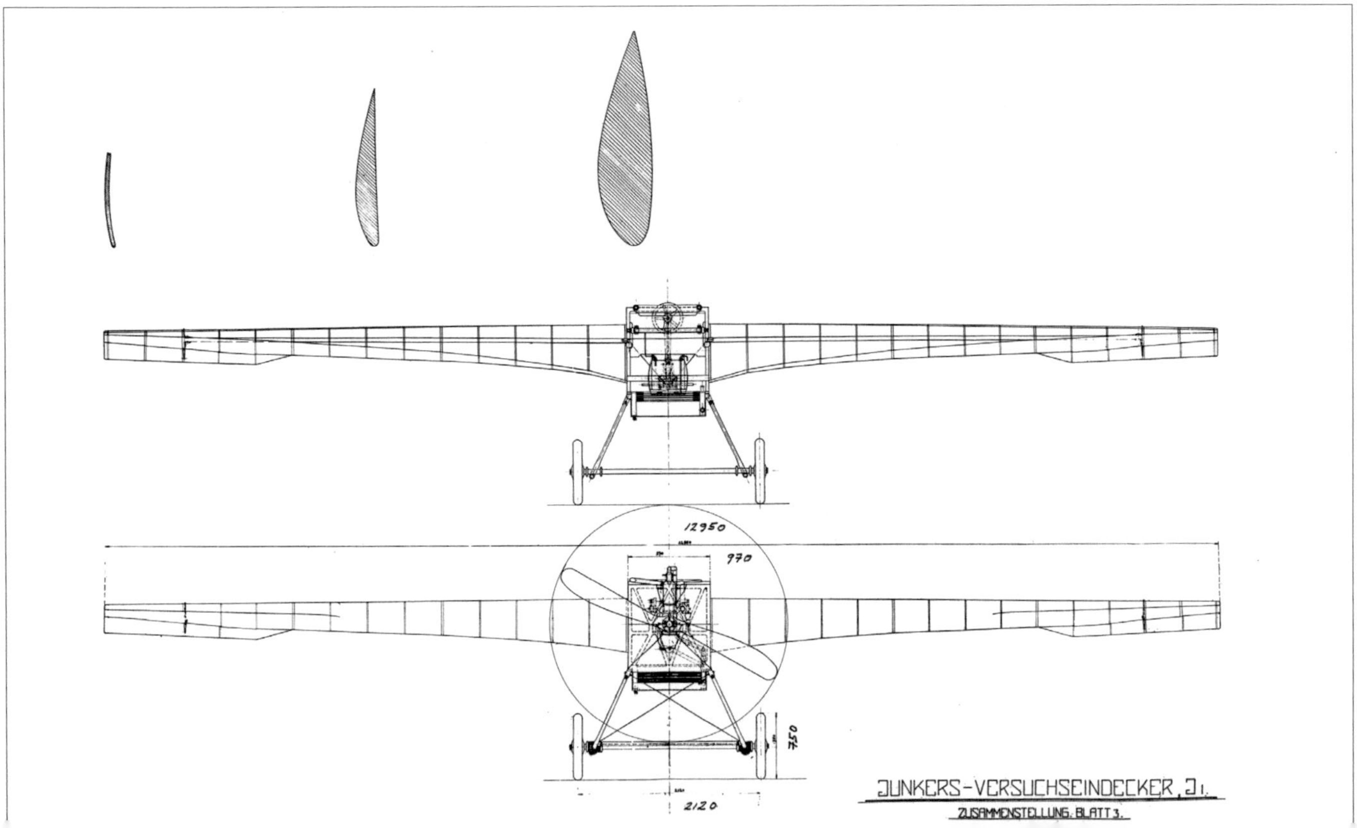

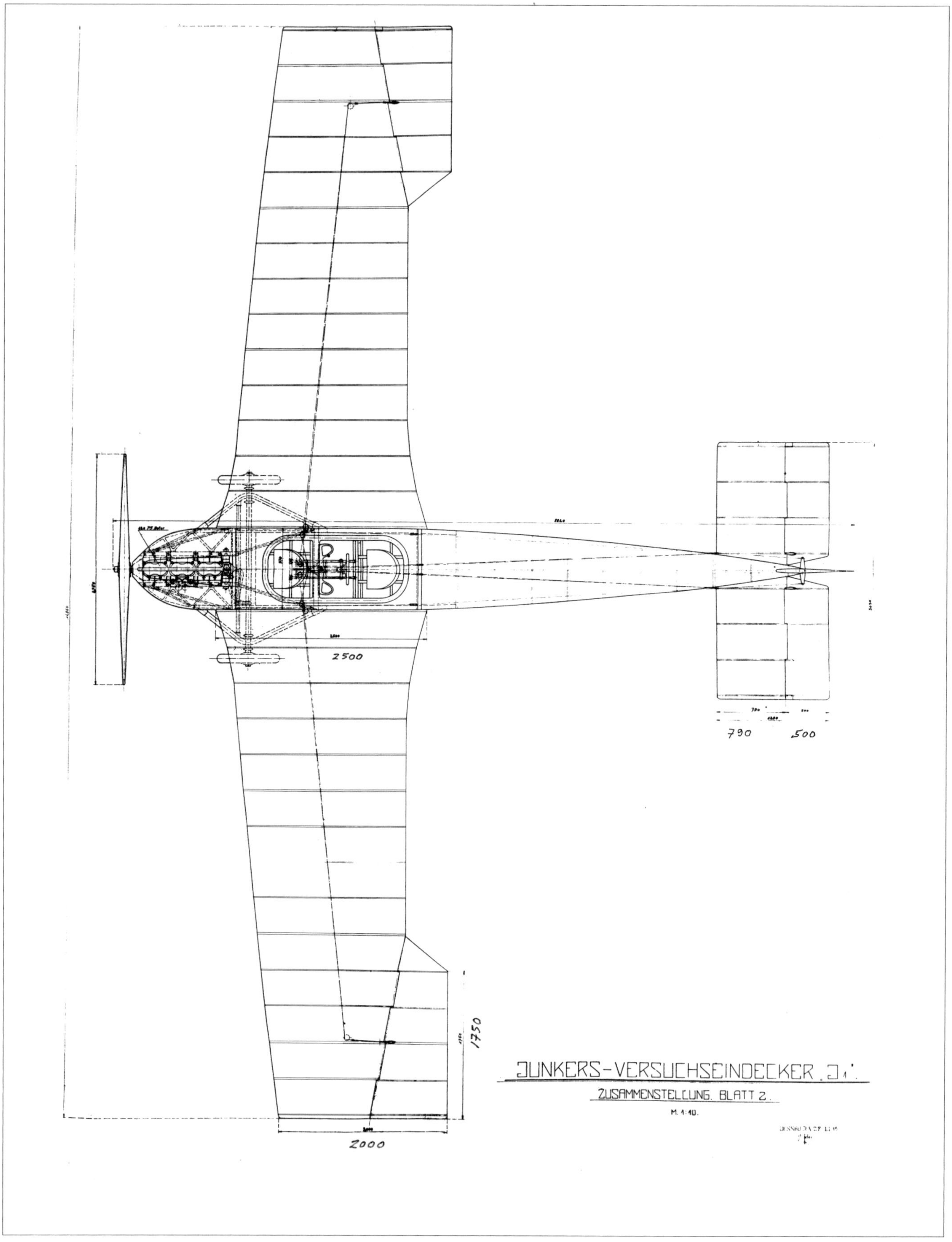
2500
790
500
1750
2000
JUNKERS-VERSUCHSEINDECKER „J 1".
ZUSAMMENSTELLUNG. BLATT 2.
M. 1:40.

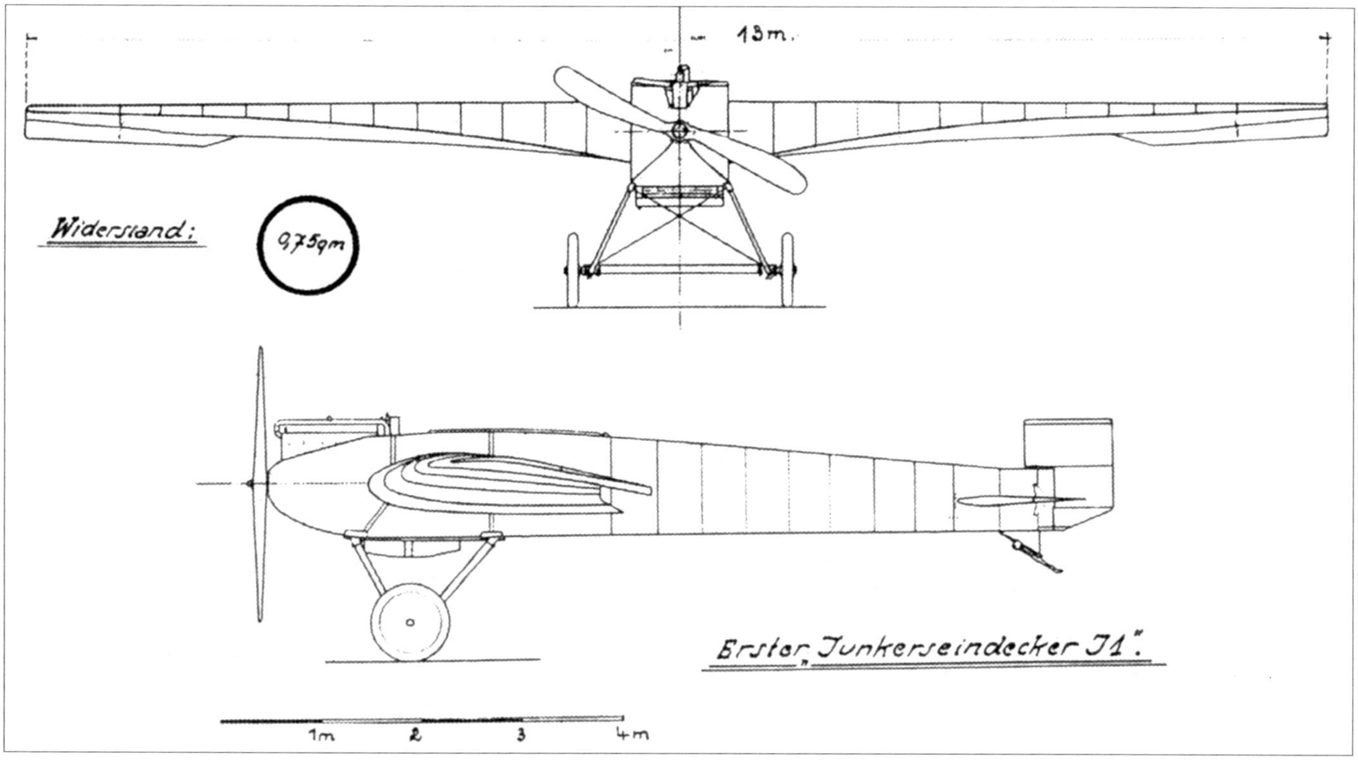

Above: These drawings of the J 1 accompanied Junkers address to the Royal Aeronautical Society in 1923.

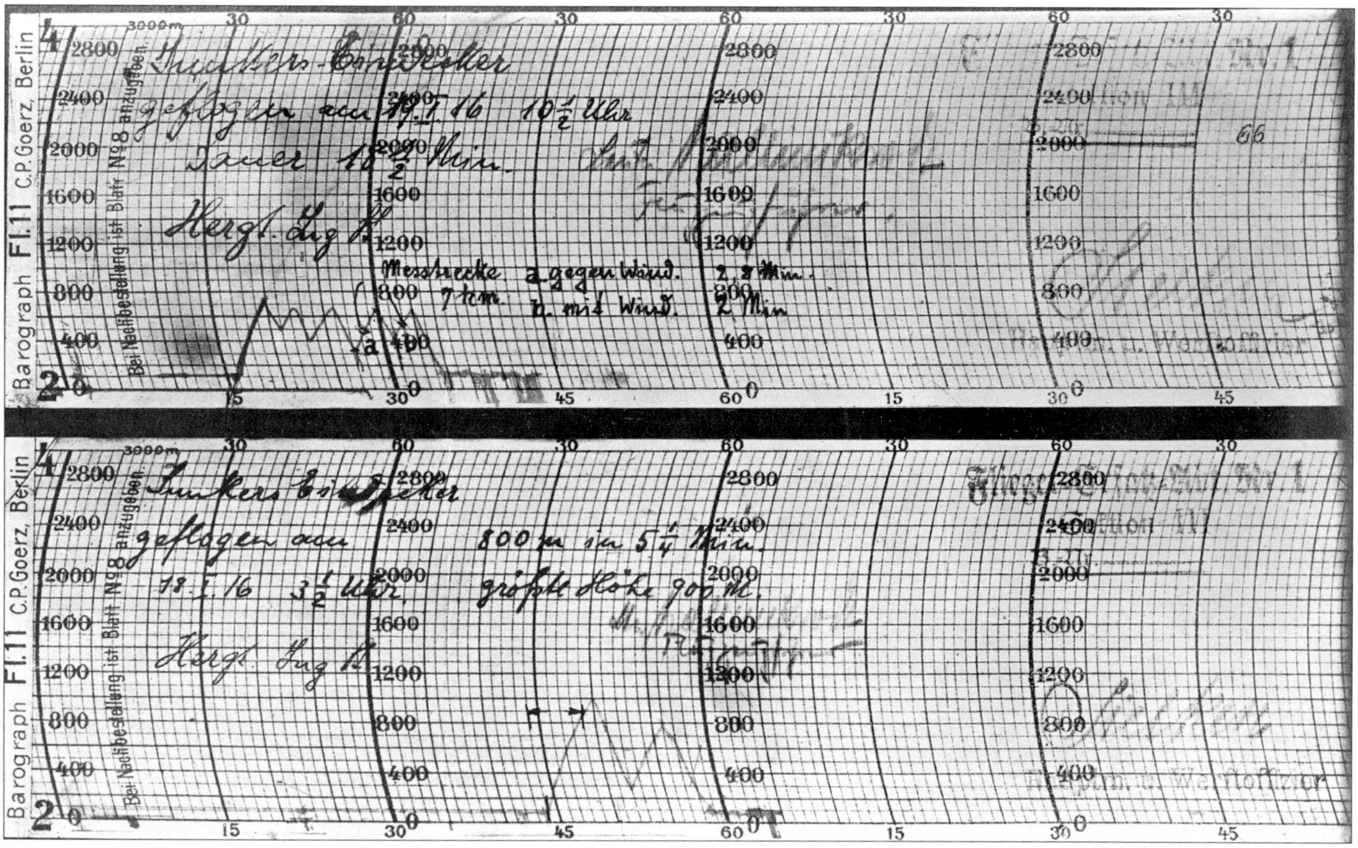

Above: Barograph recordings of the J 1.

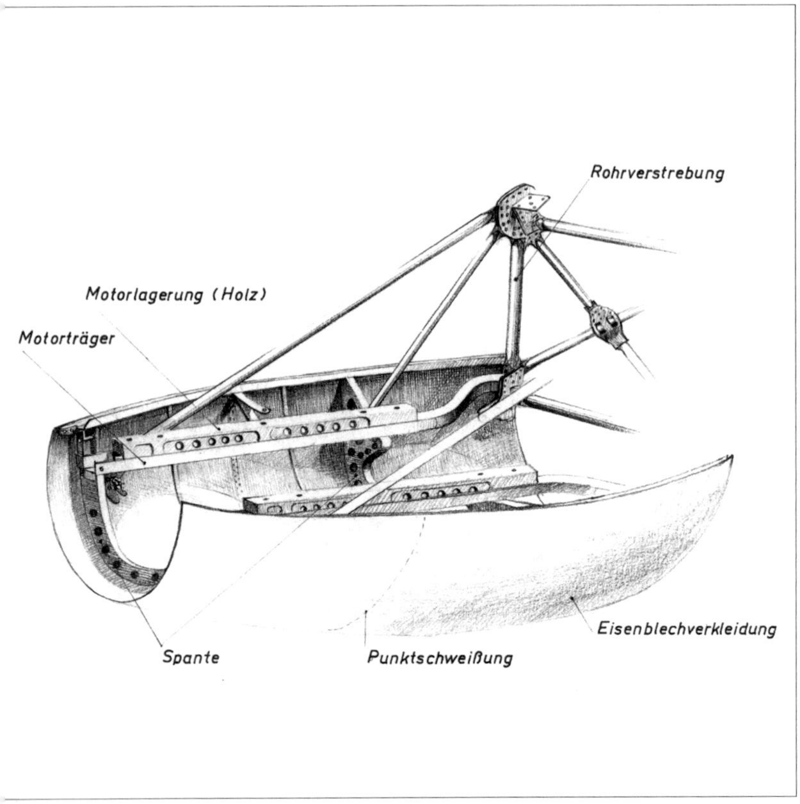

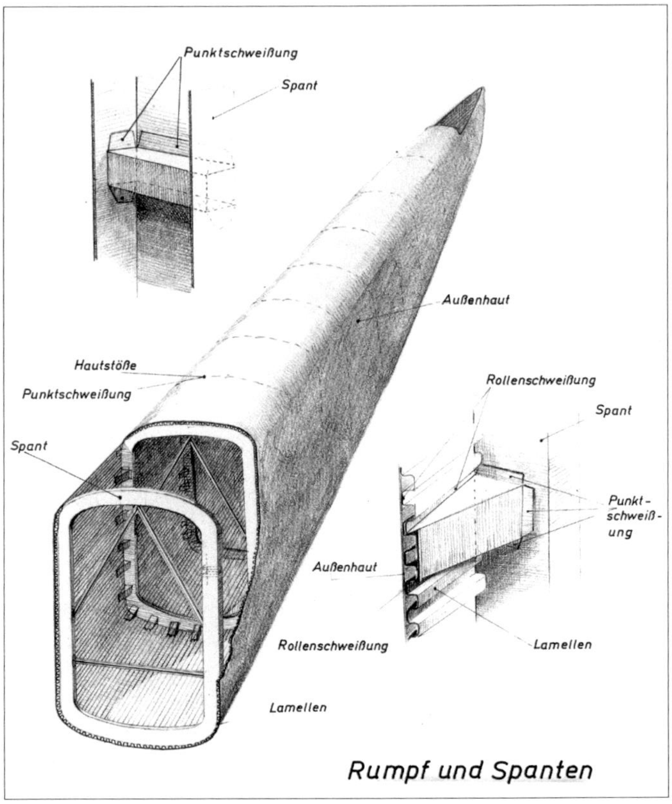

Above& Below: Details of the Junkers J 1. (via Technikmuseum "Hugo Junkers")

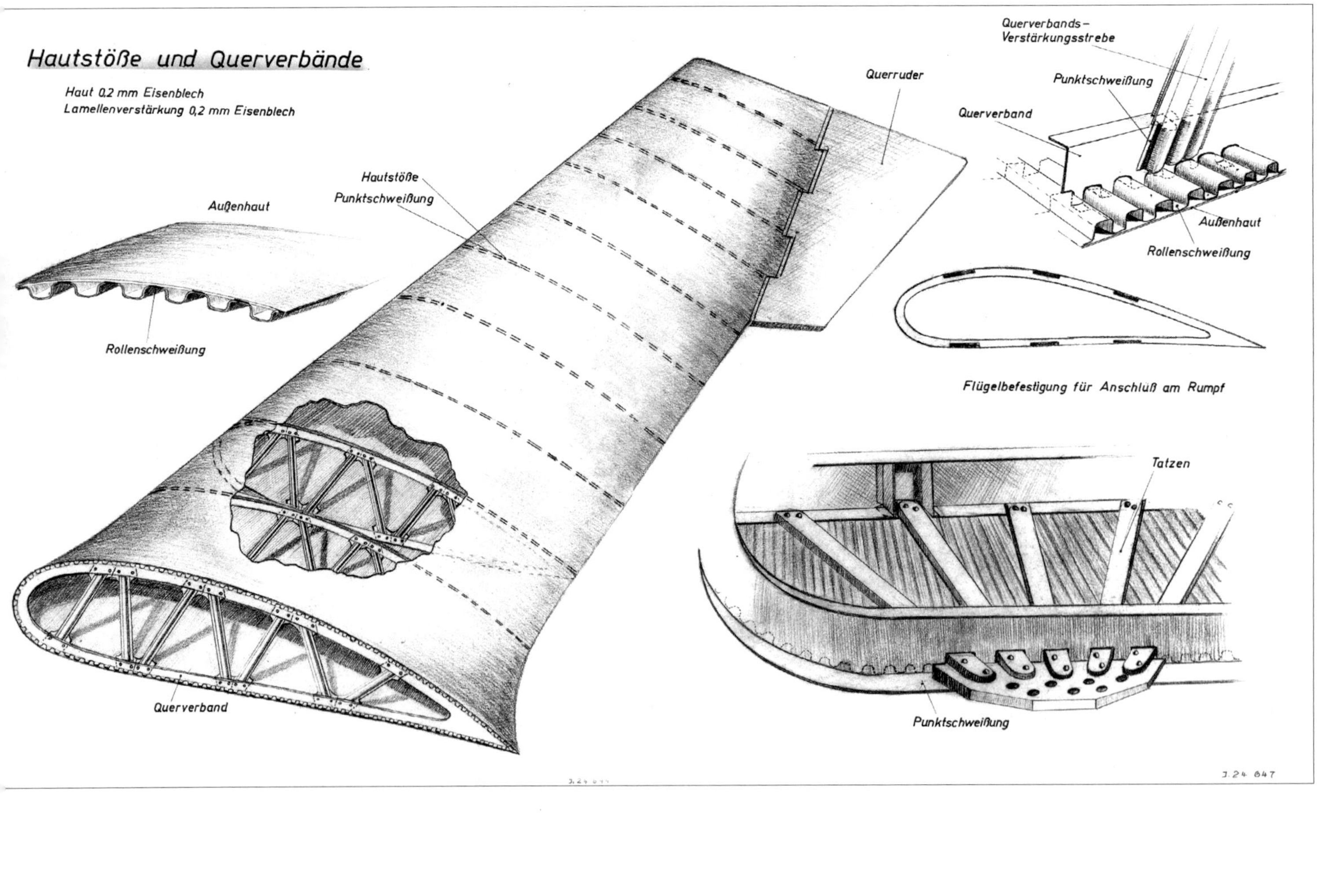

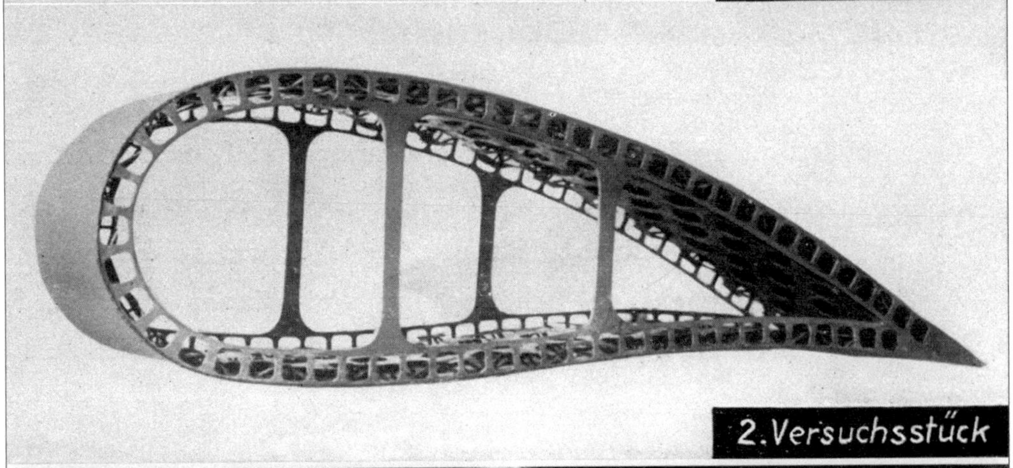

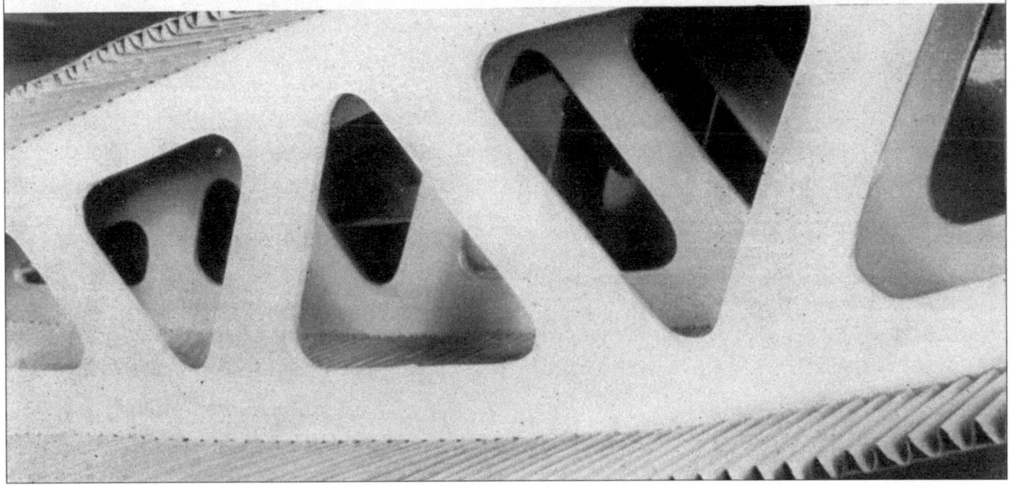

This Page: Various methods tried out in order to develop a method of fabricating metal wings using sheet steel. The corrugated inner skin is visible on the bottom experimental sections.

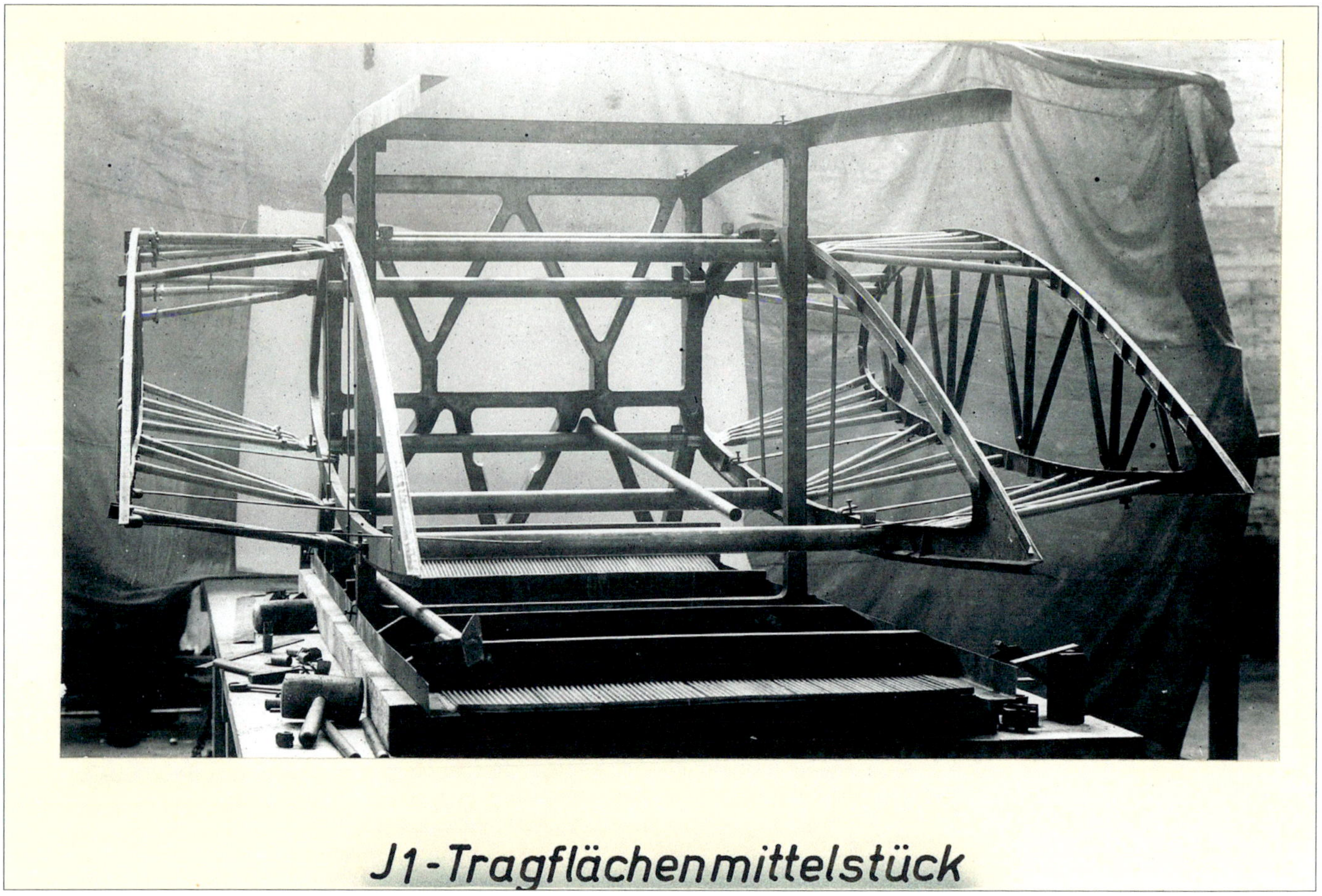

Above: The steel tube fuselage of the J 1 with the wing connections attached. Note the six attachment points for the wings to the fuselage. The fuselage suffered distortion at this point and led to the Junkers method of building the wings into the fuselage structure.

Above & Following Page: The J 1 monoplane's wing on the construction jig. The small panels were assembled into strips and joined by electro welding. The spacing of the ribs was determined by the width of the metal panels. The corrugated inner metal skin appears to have been bent into shape before welding to the outer skin.

trousered landing gear of the wind tunnel models being eliminated as a compromise for ease of construction. A tail skid with a simple iron shoe was mounted at the rear of the fuselage. The radiator for the Mercedes engine was underslung between the landing gear struts for aerodynamic reasons.

Power was supplied by a 120-hp Mercedes D.II engine mounted in the pointed nose. The radiator was mounted in an advanced box like structure under the fuselage. The pilot occupied the rear seat

while a flight test observer could occupy the forward position to record performance.[4]

The centre section of the wing was built into the middle fuselage section as part of the fuselage structure. The wings were built up similar to the rear fuselage in sections welded together such that the wing tapered from the centre section to the tips. These panels were connected to the central fuselage by means of six fittings. The wing was of Junkers patented construction. Span-wise steel tubing was tied together with a triangular truss-work. The sheet steel wing was covered by 0.2 mm steel plates, stiffened by corrugated panels welded inside the structure. The ailerons extended beyond the wing spanwise and chordwise. Building time for this machine was approximately 24,000 hours, between July and December 1915.[5] It was very difficult to obtain the thin steel sheets and then only in small sizes and quantities. As a result of this the construction of the steel machines was

Above: Metal ribs were constructed in different ways in order to come up with a light strong structure. The whole construction process was one of experimentation.

Above: Fixing the wings to the fuselage. The engine bearers were of wood. The construction area was separated from the field kitchen fabrication area by a board partition.

considerably increased. The material costs excluding engine, propeller, radiator and instruments were approximately RM 2.000.[6]

After structural testing, the J 1 was transferred to Döberitz as Junkers lacked a suitable airfield at Dessau. The first flight took place on 12 December 1915, with *Lt.* Theodor von Mallinckrodt at the controls. The *Abteilungsführe*r (Division Commander) gave strict orders to make the first hops with extreme caution, not to fly the machine but only to lift it from the ground. However, it rose up after a run of about 40 metres and on reaching a height of about two to three metres Mallinckrodt attempted to land and in doing so the left wing

Above: AEG electric welding machine as seen in the photograph below.

Left: This appears to be a gas welding machine

Above: Electric welding of a wing, 22 June 1916. The welding techniques were still in an experimental stage and sometimes stresses appeared as a big blister in the sheeting.

touched the ground and the fuselage buckled inwards at the fuselage to wing join.

Mallinkrodt reported that together

with 20 or 30 comrades of the FEA (Flieger-Ersatzabteilung) I was standing in front of the J 1. We were so filled with respectful amazement at the novelty of this design that no one would initially dare utter a word. But then came the expert criticism. No one would after all consider that it was possible that the "Tin Donkey" could fly. It

Above: A wing section in the electric welder. This appears to be a wingtip section.

Above: A view of the aircraft working area.

Left: Welding the corrugated sheet to the inside of the smooth wing skin was complicated and the Junkers workers had to gain experience with the still imperfect welding techniques.

is true that this J 1 impressed me, standing there without any struts or wires.

I volunteered to test the J 1. "You must be sick and tired of life to fly an aeroplane without struts or wires." Such were my comrades' immediate scornful questions. The warnings could not waver my resolve that a successful flight must begin a tremendous revolution in aviation. As I had reckoned with a very long take-off run, I determined the field had an unsatisfactory length and so I had the aeroplane fitted with two handles I could grab in case the machine turned over.

On the morning of 12 December 1915, I received

Above: It was difficult to weld the individual wing sections together, especially where different thickness of metal were involved. The light weight of the wing is evident in this photograph taken in June 1916.

Above: The J 1 complete at Döberitz in the *FEA* 1hangar, 12 December 1915. The elongated cockpit in front of the pilot for an observer is shown. The small size of the metal panels available to Junkers is shown by the joins of the ten fuselage sections on the rear fuselage.

permission to fly. My comrades (amongst them the then flying student Manfred von Richthofen) gathered at the long side of the field with officers of the FEA, the military authorities, and Professor Junkers with a few of his engineers. I was once more made aware of my critical task. No one could know how this iron aircraft would behave in the air. The heads of departments gave me strict instructions to make the first leaps with extreme caution.

Rolling over the field I felt the wings oscillate strongly but after only 40 metres the plane lifted off the ground. I had expected the take off distance to be at least 250 metres. The wonderful harmony with which I flew away at a low height, it is true, was very soon replaced by severe vibrations. According to the old pilot's principle "If it gets difficult up there; power off and stick forward." I prepared for landing but the left wheel struck first.

Above: The J 1 undergoing airscrew thrust measurement at Döberitz on 3 December 1915. (SDTB)

Above: Complete with military markings the J 1 ready for flight.

Above: Static test of the J 1 monoplane wing.

The asymmetrical strain caused the fuselage to be somewhat damaged.

The military authorities congratulated Professor Junkers who looked quite broken, not deigning to give me a look. I had not flown the machine with the success he expected. In contrast to Professor Junkers, the engineers were in the best of moods – it had been proven that the "Tin Donkey" could fly.[7]

"In spite of the short test, it was apparent from the whole behaviour of the machine that the flight and lifting characteristics were good." It was criticised that the stiff wing compressed the fuselage sideways as the fuselage was not constructed substantially enough.[8] The aircraft was repaired in the workshops of *F.E.A.* 1 and the wing to fuselage join strengthened.

After repairs testing continued with the first actual flight occurring on 18 January 1916, at the hands of *Gefr* Arnold of the flight school. The machine takes some 200 metres for the take-off, and Arnold made a circuit of the field at about 80 metres height. After a successful landing it was discovered that the elevator's trim tab has been set in the wrong position before the flight as the machine was thought to be tail heavy, this leading to the long take-off run and made the machine nose heavy in flight. The same day Mallinckrodt made a flight after the tab was set to the original zero position. After a short run, he climbed quickly to about 900 metres altitude. After 14½ minutes he landed but ran into a ditch bending the wheel axle. No damage to the landing gear or the wing attachment to the fuselage could be found. Mallinckrodt reported that the aircraft was no longer nose heavy and is perfectly normal in the air.

The following day at about 10.30 a.m., Mallinckrodt undertook speed measurements with and against the wind. The pilot was instructed to fly along the straight line between the *FEA* workshops and the church at the village of Fahrland as the wind was blowing in this direction. The wind speed was about 3 m/s. The machine took off quickly and

Above: The J 1 is the object of interest at on 18 January 1916, at Döberitz. The concept of an all-metal aircraft was radical at the time. (SDTB)

without difficulty.

After a few turns, climbs and glides, the machine flew over the (FEA) *workshops at about 200 metres height towards Fahrland. The pilot had instructions to clock the time to cover the distance from the workshops to Fahrland. A barograph was used for control. During the speed test, the machine climbed almost 300 metres over a distance of 7 kilometres. The time recorded by stop-watch was 2 minutes 40 seconds while the barograph showed 2.8 minutes. Beyond Fahrland the machine made a turn, gliding down 200 metres. For the return flight over the same distance the machine recorded a time of 2 minutes 5 seconds in the tail wind, while the barograph showed 2 minutes. The landing from 550 metres height was easy.*

To have a comparison of the speed and to check the distance really flown, the same distance was flown by an Albatros DD with a 120 hp Benz. With a tail wind of 4 m/s the flight time from Fahrland to the workshop was slightly over 4 minutes. The recalculation gives the monoplane a speed of ... about 170 km measured with and against the wind. The weight of the machine for the flight, including fuel and oil, was 1,010 kg. [9]

Oswald Bôlcke, who inspected the machine, around 23 May 1916, proclaimed that "if this machine reaches 170 km/h it is the fastest aeroplane so far!"[10]

The J 1 proved to have a reasonable performance despite being heavier than contemporary fighters at a weight of 1,010 kg. The aircraft achieved 170 km/h which compared favourably with the wood and fabric fighters of the time. The J 1's other aeronautical properties were not very good as it climbed badly and was sluggish on turns which was to be expected considering its weight.

The J 1 was not liked by the ground crews as it presented an entirely new set of problems that had to be solved. It was criticised that the stiff wing compressed the fuselage sideways as the fuselage was not constructed substantially enough.[11] Despite this, as an experimental machine, it was a success. This led to the *Inspektion der Fiegertruppen* (*Idflieg* -Inspectorate of Flying Troops) granting a contract on 31 January 1916, for six improved single-seat fighter versions of the design, the J 2.

Above: Running up the motor of the J.I at Döberitz prior to the first taxi test by *Lt.* Theodor Mallinkrodt on 12 December 1915.

Above & Following Page: *Erster Metalleindecker J I (1913) mit freitragenden Flűln.* The J 1 on display in the Munich Technical Museum after the end of the war. The date (1913) was incorrect. A Fokker D.VII and what appears to be an Etrich Taube are also on display.

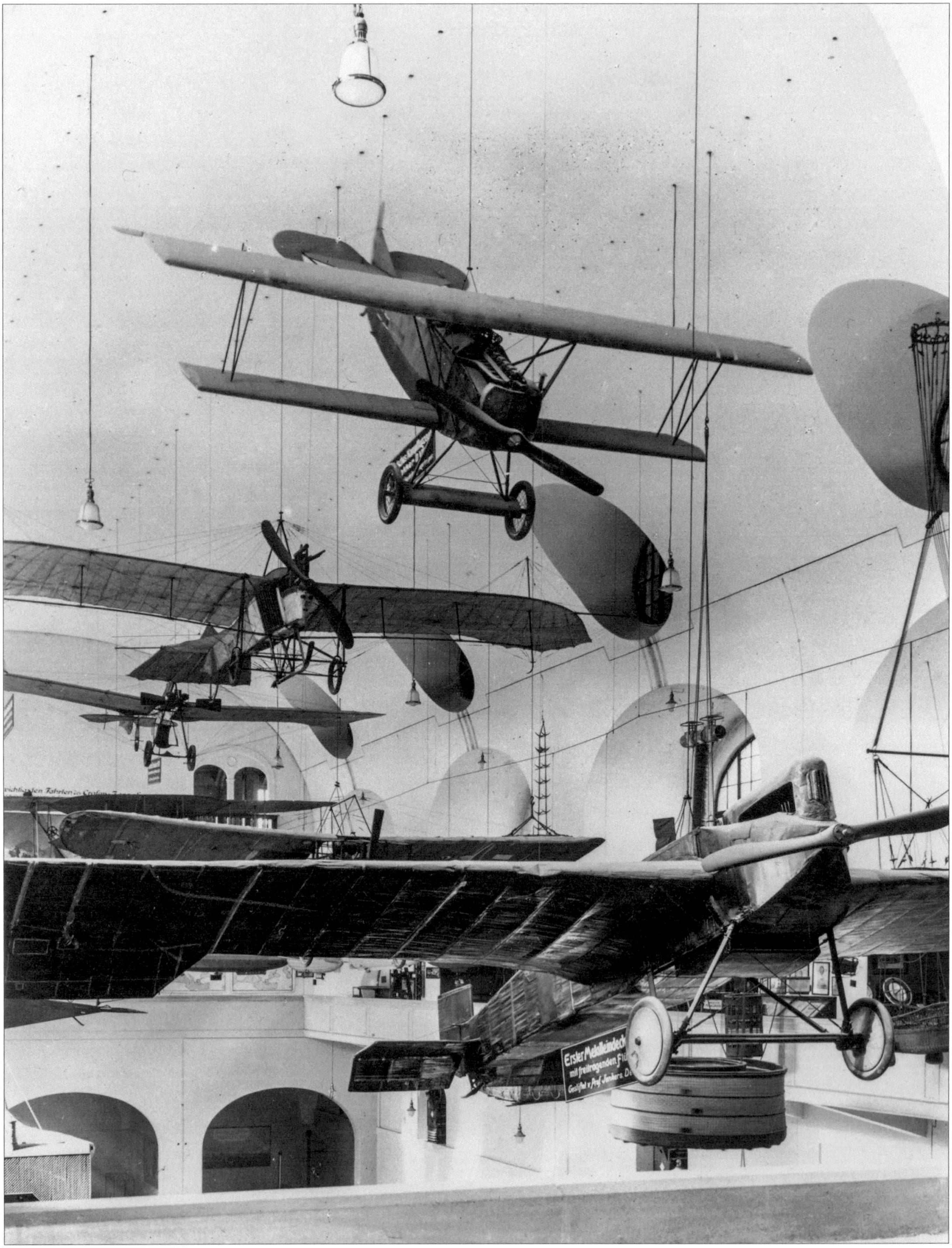
Erster
mit freitragenden
Gestiftet v. Prof. Junkers.

Above: Volunteers working on the reproduction J 1 at the Technikmuseum "Hugo Junkers", Dessau, July 2017. The cockpit section with engine bearers is almost complete.

Below: Wing ribs for the J 1 reproduction.

Above: The wing skin of the J 1 comprised the outer metal sheet steel with an inner corrugated skin. Here the volunteer is assessing how the skin will be attached to the wing structure.

The J 1 was then displayed at the Junkers Educational Exhibition at Dessau. From 1926 the machine was given to the Deutsches Museum in Munich and is presumed destroyed in 1944 during the Allied air raids of World War II along with other priceless aeronautical relics.

The *Junkers Technikmuseum* in Dessau has started to build a reproduction J 1. A Crowdfunding campaign in May 2016 raised 42,000 Euros, 40% over the goal.

Endnotes Chapter 2

1. The official *Zentrale Abnahme Kommission (ZAK* – a part of *Idflieg)* report, *Geschichte der Deutschen Flugzeugindustrie* (*History of German Aircraft Industry*), produced by *ZAK* in 1918, states that Junkers started the preliminary layout for the metal cantilever wing monoplane in May 1914. He recognised the advantages of the cantilever wing and metal was preferred to wood because of its consistent strength and good resistance to fire. Translation by Roland Jahn. Hereunder referred to as *ZAK Report.*
2. According to the *Ganzmetallflugzeug J1u.J2 Stahlbauweise* the sheets were from 0.2 to 0.5 mm thick. Most other sources state 0.2 mm only.
3. H. Junkers, quoted by H. Woodman. "The Tin Donkeys of Dessau," *Airfix Magazine*, May 1983, P.390.
4. There is no record of the machine flying with two persons aboard and Peter M. Grosz was of the opinion that it was unlikely that the machine was ever flown with an occupant in the second seat.
5. Wagner, W. Gives September 1915 as the start of the construction of the J 1, apparently relying on the recollections of Otto Seifert. Wagner, W. *Hugo Junkers Pionier der Luftfahrt – seine Flugzeuge,* Bernhard & Graefe, Germany, 1969.
6. *Ganzmetallflugzeug J1u.J2 Stahlbauweise.* Junkers Museum, Dessau.
7. "Über den Blechesel und fliegenden Tank zur Junkers J 9 (D.I). "*Aerokurier* 3/1973. P.221.
8. H Heinrich to A.R. Weyl, 05.11.1925. Note in P.M. Grosz Collection Box 487.

Specifications J 1					
Source	**1**	**2**	**3**	**4**	**5**
Dimensions in m					
Length	13	12.95	12.95	12.95	12.95
Span		7.43	8.62	8.60	8.62
Height		3.13		3.11	3.11
Areas in m^2	24	24.64	24		24.60
Weights in Kg					
Empty		900			900
Fuel					58
Oil					5
Crew					80
Useful Load					37
Total load					180
Loaded	1,010	1,010	1,010	1,080	1,080
Speed in km/h					
Max	170	160	170	170	170
Cruising					160

Sources:
1. Junkers, H. "Metal Aircraft Construction."
2. Grey & Thetford, *German Aircraft of the First World War.*
3. P.M. Grosz Collection Box 484.
4. *Ganzmetallflugzeug J1u.J2 Stahlbauweise.*
5. Wagner, W. (Junkers *Pionier der Luftfahrt – seine Flugzeuge*, Bernhard & Graefe, Germany, 1969.)

9. Report from FEA 1, Döberitz, to *Idflieg*, Berlin. 20.01.1916. Copy via R. Zankl
10. *Auszug aus dem Protokoll über Besch Bôlcke in der Fabrik Junkers & Co., Dessau, am 23. Mai 1916.* via B. Schmäling.
11. H. Heinrich to A.R. Weyl, 05.11.1925. Quoted in P.M. Grosz Collection Box 484.

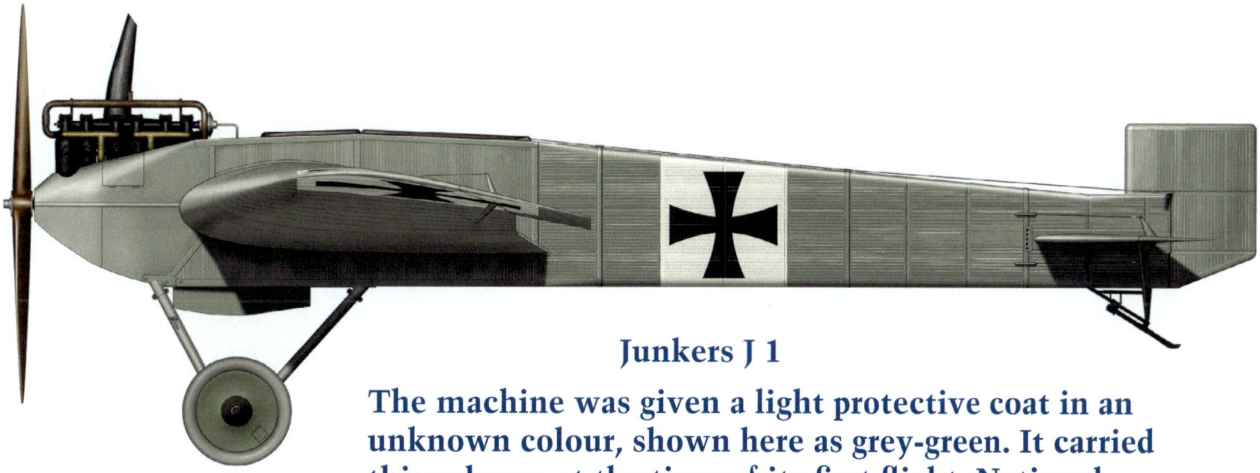

Junkers J 1

The machine was given a light protective coat in an unknown colour, shown here as grey-green. It carried this scheme at the time of its first flight. National markings were carried on the top and bottom surfaces of the wings but were absent from the rudder.

3. Junkers J 2 (Military Type E) – The First All-Metal Fighter

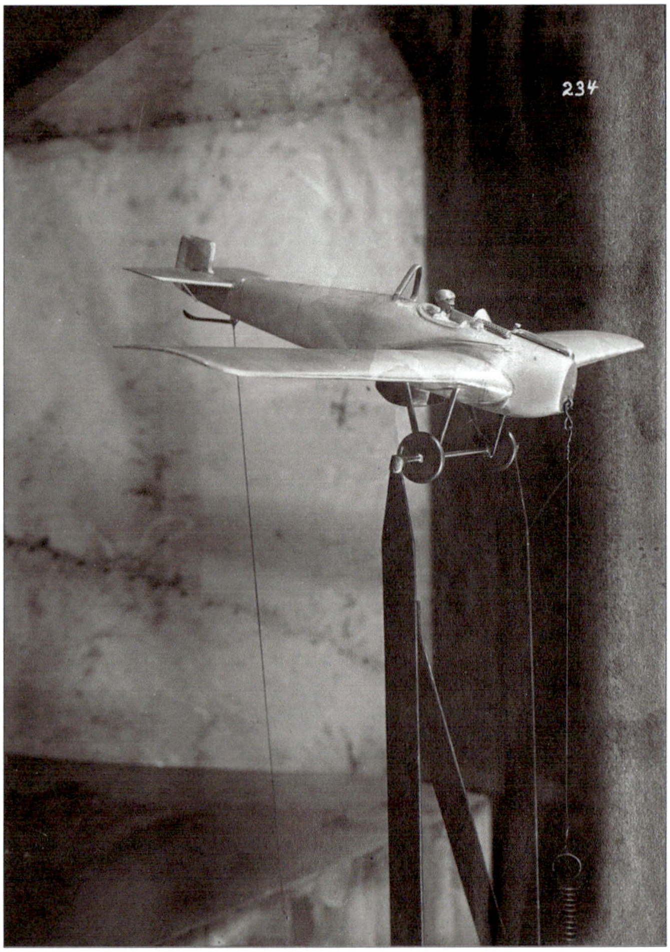

Above: Wind-tunnel models of the J 2 under test. The model includes a pilot and machine gun. (Junkers Album via R Zankl)

With the "technology demonstrator" J 1 having proved the concept, *Idflieg* ordered six all-metal single-seat fighter aeroplanes from Junkers in early 1916; however, Junkers had already completed the first. The machines were to have the same Mercedes engine as the J 1 and were to be armed with a single synchronised LMG 08/15 machine gun.

The J 2 was structurally the same as the J 1 but of smaller dimensions with a strengthened undercarriage. The fuselage was constructed the same as the J 1 but was more-rounded. A pylon was fitted behind the pilot's cockpit to prevent injury in the event of a roll over and the120-hp water-cooled Mercedes D.II engine was completely enclosed. Cooling was provided by an underslung radiator. The undercarriage was high, higher than that of the J 1, and the legs were connected to the wing structure rather than the fuselage as in the J 1. The engine mount, central wing section and cockpit were built as the one integral unit. Again, the balanced rudder was fitted without a fin. Unfortunately, while an elegant looking aircraft, the J 2 was almost the same weight as the J 1, and with a higher wing loading its performance was poorer than that of the J 1. The six machines were accepted by *Idflieg* with the serials E.250/16 to E.255/16.[1] The first two J 2 monoplanes were to be fitted to carry a single synchronised Maxim LMG 08/15 machine gun.

E.250/16 was sent to Adlershof for sand load testing on 2 July 1916. Upon obtaining satisfactory results E.251/16 was dispatched to Döberitz for flight testing, making it first flight on 11 July 1916.

Above: Forward fuselage framework of the J 2 looking forward from the rear. The two fuel tanks in the wing roots may be seen.

While the test pilots were enthusiastic about the J 2 with its good flight characteristics, it was grossly overweight with poor climbing characteristics. The machine reached 180 km/h under the hands of test pilot *Ltn* Max Schade. In an effort to improve performance the remaining machines were modified. The engine was replaced with a 160-hp Mercedes D.III, overall length was increased slightly, the wing span was increased by 700 mm while wing area was reduced by 84 cm^2, long span short-chord ailerons were fitted and the forward fuselage contours revised. It is probable that these modifications were necessary due to the prototype being somewhat tail heavy and directionally unstable. The four machines under construction were to incorporate these modifications.

In its final form, the J 2 was some 15 km/h faster than contemporary fighters[2] but the ever increasing demands of the Front meant that its performance fell short of what was now required, particularly

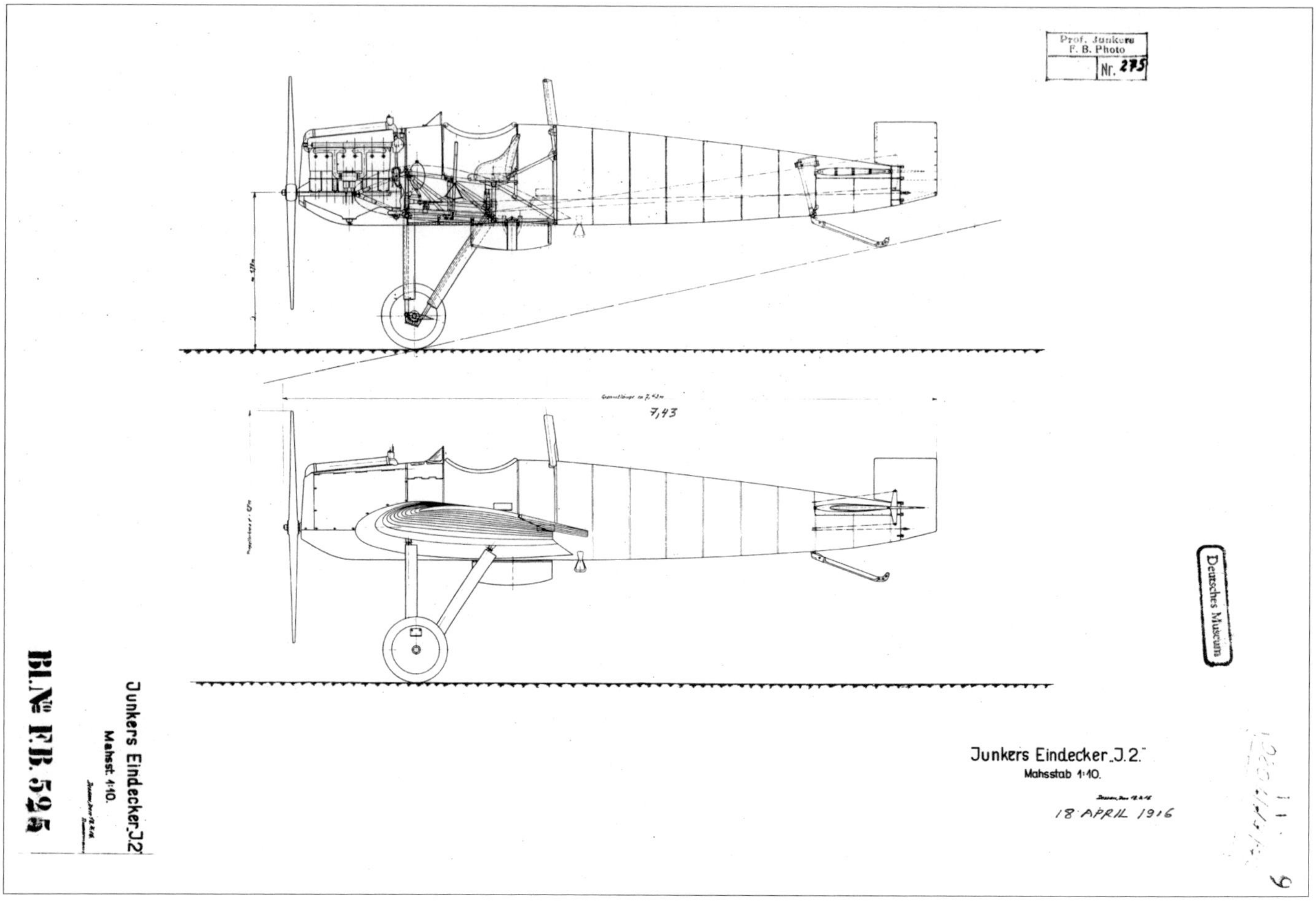

Above & Facing Page: Junkers drawings of the J 2 *Eindecker* 18 April 1916. The wing to fuselage structure is well depicted and should be compared with the photographs.

with respect to climb. Junkers himself complained as to how much the specifications had changed from the placing of the contract to the testing of the machines. The problem for the J 2 was mainly due to its weight. After E.252/16 was fatally crashed from a spin by *Ltn* Schade, on 20 September 1916[3], *Idflieg* apparently lost interest in the J 2. Schade had attempted a slow flight over Dessau. The remarkable J 2 had met all the original military specifications but the pace of air fighting had been such that the J 2 would not be able to meet the requirements of the Front. Junkers concluded that the steel aeroplane was not the way to achieve the all-metal aeroplane. What was wanted was a low weight to power ratio. The J 1 and J 2 had been designed strong for safety but this had resulted in a machine that was too heavy. Junkers would have to go back to the beginning and start again.

Anthony Fokker flew E.252/16 on 22 December 1916. It is worth noting that his interest in thick wooden wings began around this time. Junkers was convinced that Fokker had adopted the thick aerofoil from his design and Fokker never denied it in his autobiography. Junkers sued Fokker post-war for using his patented wing design and won.

Just before the war Junkers began to consider his financial position. Working in his design facility he pursued his theories but the machines required for testing and the tests themselves were expensive and cost him dearly. His competitors fought his patents, in particular AEG and MAN. While the income from patent licence fees was good and increased yearly, Junkers actually lived "hand to mouth". Hitherto he never considered the need to safeguard his companies as he worked unfailingly towards his research goals. He now came to the conclusion that

I have hitherto concentrated everything to the development of new things in calm and tranquillity, undisturbed by the outside world. For this reason, I have avoided contact with the financial communities of the outside world.

Junkers noted that quality was not a guarantee of success. Second-rate articles took the place of first-rate ones when they were supported by

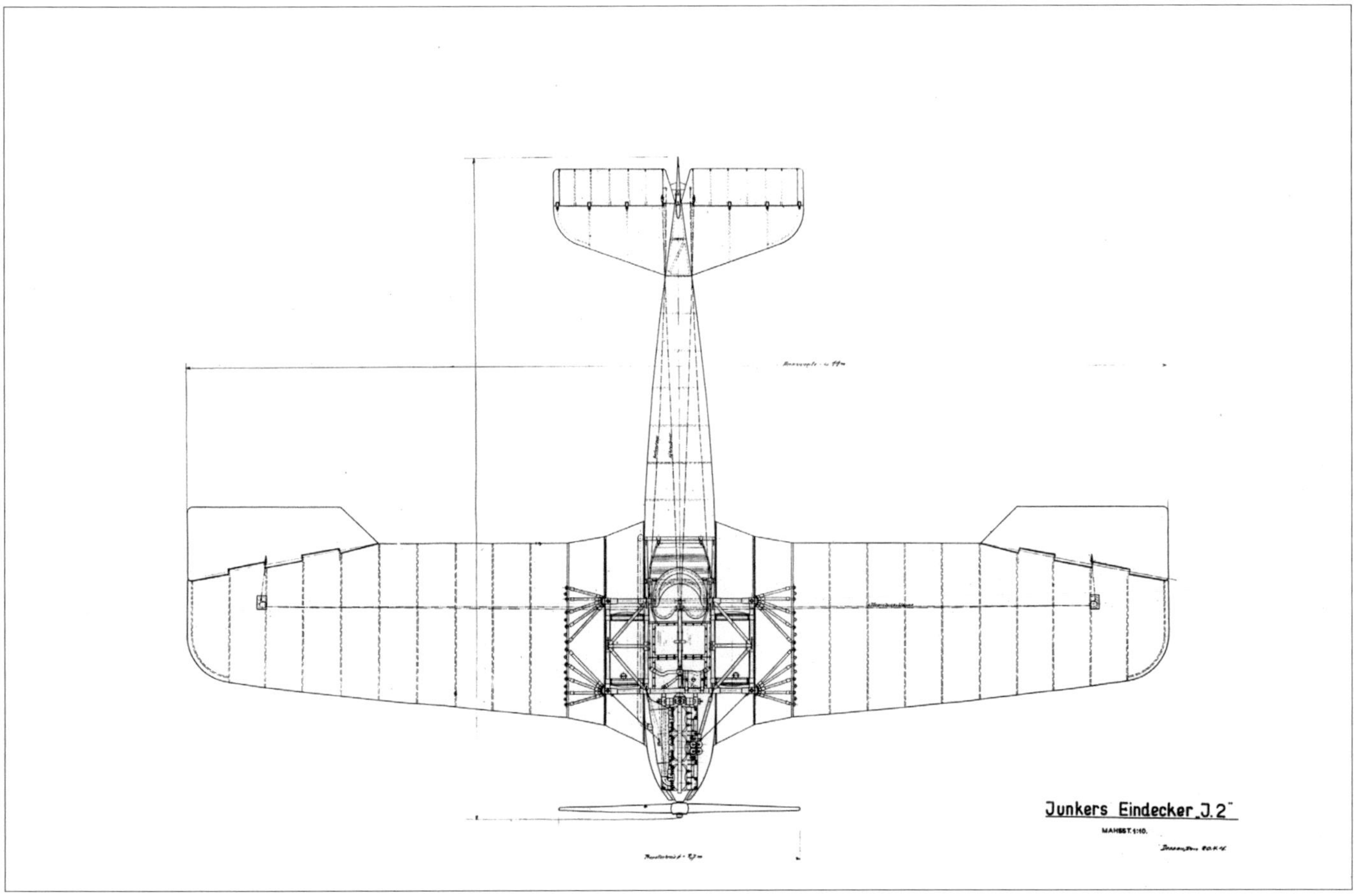
Junkers Eindecker „J.2"
MAHSST. 1:10.

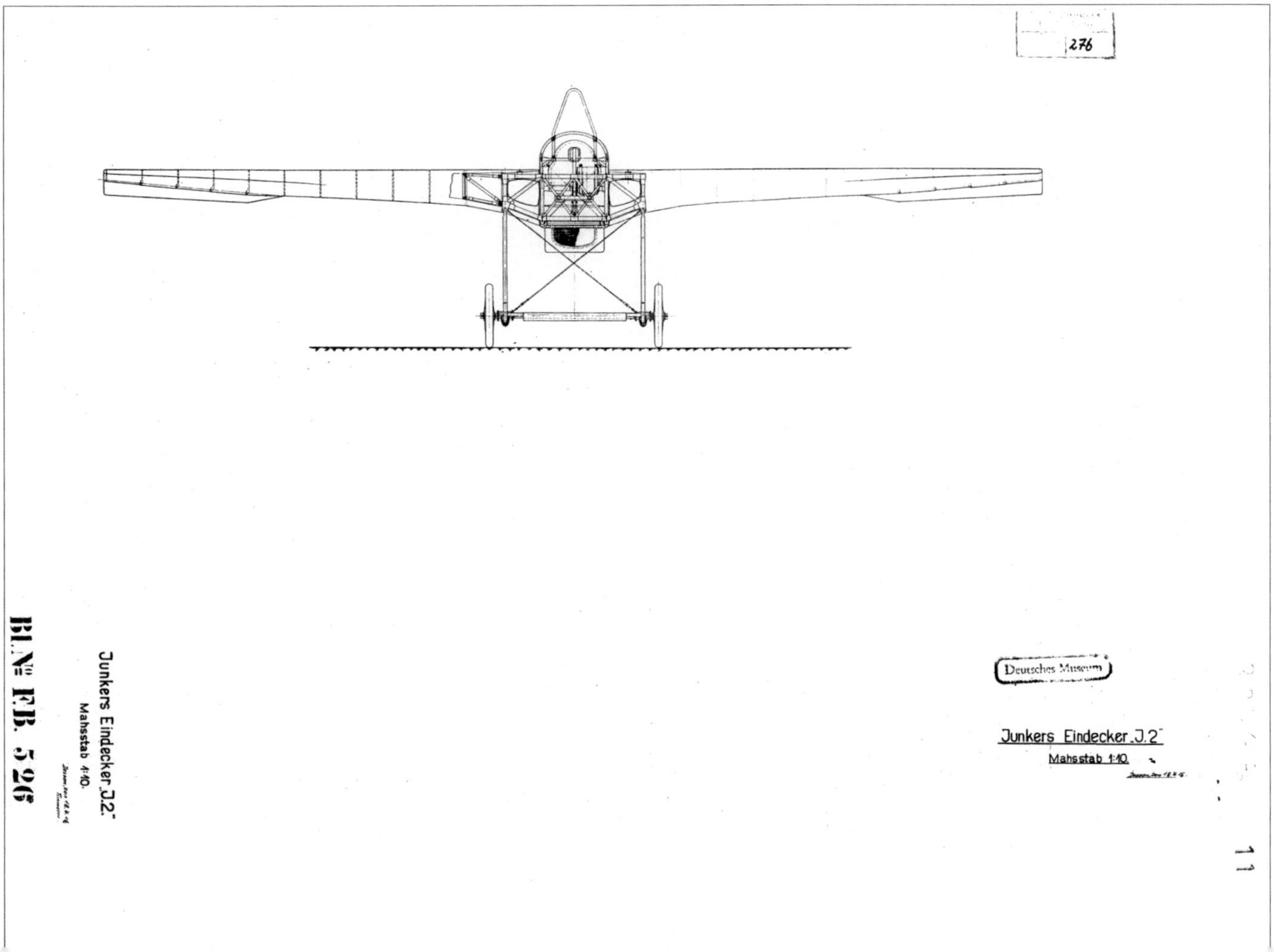
276
Deutsches Museum
Junkers Eindecker „J.2"
Mahsstab 1:10.
Bl. № FB. 526
Junkers Eindecker „J.2"
Mahsstab 1:10.

Above & Below: Engine installation on a J 2. Note the modern underslung radiator fairing.

Above: The rear fuselage section has been mated to the front section but there is no roll bar on the machine.

Below: Assembled fuselage with roll bar. 10 May 1916.

Above: The finished fuselage of E.251/16 before the wing mating. Junker's photograph 243 and 244 dated 30 June 1916. The colour is unknown but light blue has been suggested. The semi-monocoque fuselage, underslung radiator and fully cowled engine would not look out of place in the 1920s. Unfortunately, the long, stalky undercarriage spoils the lines of the machine. A single synchronised machine gun was the proposed armament. A new feature of the design was the integration of the wing centre-section with the fuselage. On the open engine panel the internal corrugated backing strip can be seen. (SDTB)

Above: The first J 2 on the tarmac ready for testing at Aldershof in June/July 1916. Power was supplied by a 120-hp Mercedes D.III and the machine has the original small rudder. (SDTB)

Above: This view emphasises the clean lines of the Junkers machine. The exhaust pipe is carried over the wing root for discharge. Note the large airship hangar in the background. The single LMG 08 machine gun is mounted. E.251/16 was the only J 2 to be armed. The photograph was taken in July 1916 at Aldershof. (AHT AL0459-212)

Above: Rear view; compare the tailplane with that of 252/16 on P.65. (SDTB)

Above: Note the underwing serials on J 2 E.251/16

Above: E.251/16 lets out a stream of exhaust smoke while running up. There are at least five ground crew on the port side of the machine.

Specifications J 2				
Source	**1**	**2**	**3**	**4**
Dimensions in m				
Length	11.70	11.00	11.00	11.00
Span	7.43	7.3	7.30	7.30
Height	3.13	3.13	3.13	3.13
Areas in m2	19.00	19.84		19.84
Weights in Kg				
Empty	1.018	1,018		863
Fuel				58
Oil				5
Crew				80
Useful Load				52
Total load				195
Loaded	1,165	1,160	1,060	1,058
Speed in km/h				
Max	200	145	205	170
Cruising				160

Source:
1. Green & Swanborough.
2. Gray & Thetford.
3. *Ganzmetallflugzeug J1u.J2 Stahlbauweise.*
4. Wagner, R.

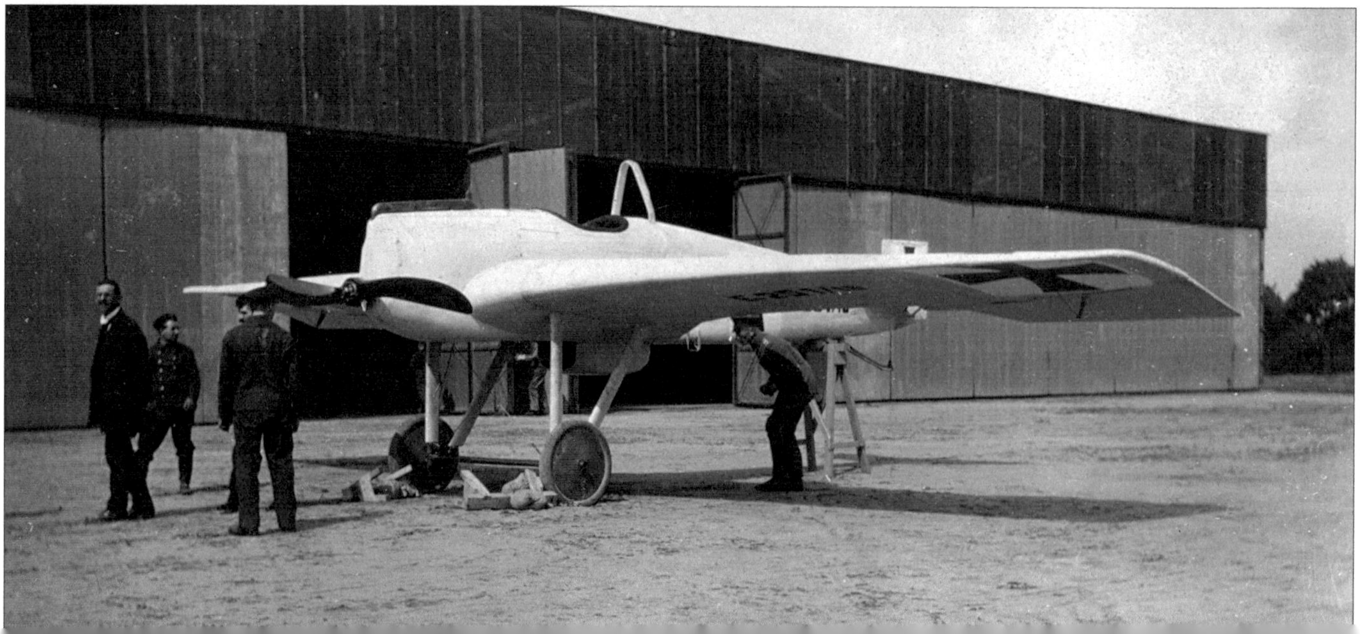

Facing Page & Above: More views of E.251/16. The serials are carried under the wings of E.251/16, possibly the first time on a German military aircraft. (Junkers Album via R Zankl)

entrenched interests. Junkers realised that he would have to promote his activities, obtain the services of "financially educated and thinking colleagues in the view of securing customers...In short, we must strive to develop and preserve more personal relationships." (Junkers notebook No.13, entry for 14 June 1914.)

The start of the war led to more problems. The licence fees from abroad stopped. Machinery for Russia was almost ready for export and this was stopped. His workforce was called up, the factory manager of Junkers & Co, Schleissing, was to report for military duty immediately. His successor, Paul Spalek, succeeded in obtaining a war contract for 2,000 field kitchens, but this was put on hold. The firm redirects its efforts to food wagons, cooking utensils, sterilization appliances, and also Junkers takes up building experimental aeroplanes in metal!

His *Aachen Versuchsanstalt für Ölmotoren* (Diesel engine test institute) was fully occupied with orders for the military. Changes also take place in the research area as well as the business areas of the companies. His big ship's diesel engines had suffered problems with German ship owners mainly due to overheating and lack of experience with the operation of these engines. The English, Russian and French firms that held Junkers licences had more success with these engines but this cooperation now ended. The Army rejected his request for financial assistance for the continual development of heavy oil engines. Further development of these large ship's engines was no longer possible during the war. One bright point was the upholding of his patents by the Supreme Court.

On 15 February 1915, the *Junkers-Werk Hauptbüro* (Junkers Works Main Office) was formed at Dessau. Junkers continued to lead the group of companies from here until his forced retirement by the Nazi regime in 1933.

The following accounts were written by a passionate Professor Junkers to the *Königliche Preussische Kriegsministerium* (Royal Prussian War Ministry) on 30 August 1917, wherein he pleaded for support from *Idflieg* to let him carry on with his system to produce the best types of superior aeroplanes. The weight problem seems to be the main stumbling block at this point in his evolution of producing an all-metal aeroplane. He does not name his detractors but it is evident there were many who did not favour his proposals. He was financially stressed and wanted financial help. To his covering letter he added three enclosures:

My Forschungsinstitut, *which was placed at the disposal of the Army administration, has created a new aircraft manufacturing system, based on its research findings, which have been fully confirmed and have shown important flight and military advantages.*

Apart from its high speed, great manoeuvrability and stability, it has important military advantages. They are:

1. *Invulnerable wings and control systems which may be shot through at many places without failing. There is no single point where a single shot can cause a complete failure of the aircraft which can occur in present-time aircraft when a wing spar it hit by a bullet.*
2. *My aeroplanes, entirely manufactured of metal, are fireproof and completely weather resistant.*
3. *The character of the metal design and*

Above: The wing attachments of the J 2.

Below: Attaching the wings of E.231/16 to the fuselage. This machine has the enlarged rudder. The light painted wingless and engineless fuselage of E251/16 rests behind. Junker's photograph dated 29 September 1916.

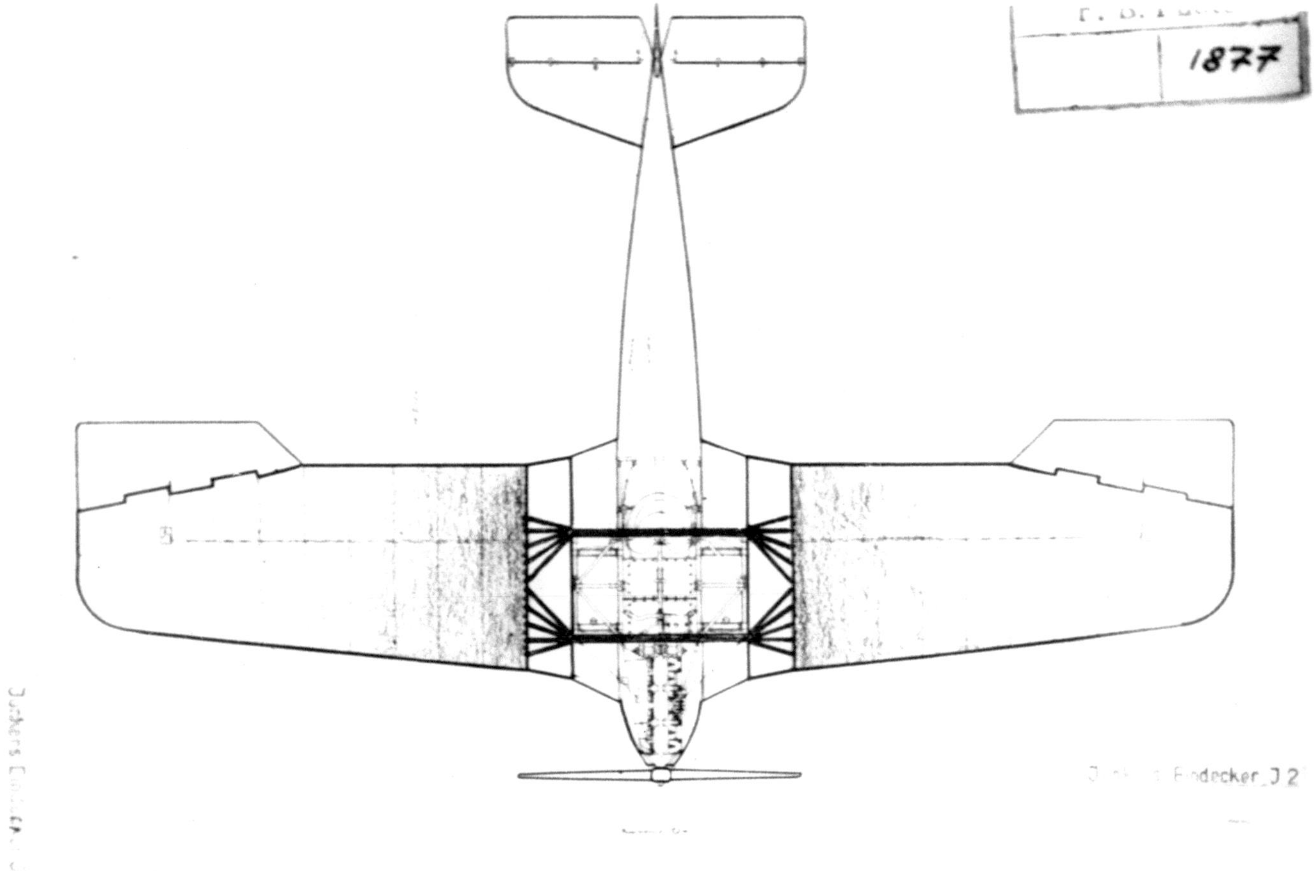

Above: Junker's factory drawing of the J 2 monoplane showing the wing attachments. (SDTB)

Above: E.252/16 running up during flight evaluation at Aldershof in Autumn 1916. The machine has a dark colour upper surfaces. The serial is carried under the wings but in small characters.

Above: *Dip Ing* Franz Brandenburg is in the suit posing with *Dr Ing* Hans Steudel in the cockpit of E.252/16, Max Schade at the left and ground crew. Schade died when he crashed in this machine on 23 September 1916. The rudder cross, on the enlarged rudder, is on a white ground that shows different from the under-surface colour, suggesting the use of a light blue for the lower surfaces. (SDTB)

Below: Front view of E.252/16. Junker's photograph No. 332, dated 20 September 1916. The instrument mounted on the port wing was probably a pitot type measuring device. E.252/16 had a 160-hp Mercedes D.III instead of the 120-hp of E.251/16, and a wing of greater area. (SDTB)

Above & Below: E.253/16, the definitive form of the J 2. Note the stencilling on the wing root. The heavily retouched photograph shows what could be done to a photograph to emphasize different features of a machine. (AHT AL0459-213)

construction make possible a cheap series production using mainly untrained workers, also female), as soon as necessary workshop equipment is available. (Note: The metal processes required a higher level of skill as they had to be developed and there was not a cadre of such workers to call on). *In spite of support from all relevant Army administration authorities, e.g. by assigning a test order for six aeroplanes, I had, in order to operate the* Forschungsinstitut, *to commence production, the design development and to establish new workshops, to spend very substantial amounts of money. To obtain these personal resources I had to use my credit to the utmost and therefore I now ask the Königliche Preussische Kriegsministerium to come to my assistance in the following suggested form. My request should now be even more well-grounded as the military usefulness of my aeroplane system has been proven by the loading and flying tests so far made.*

Support which might come to the fore:

1. *Remuneration of part of the costs I have had for the aerodynamic research facility (wind tunnel).*
2. *Awarding of further orders.*
3. *Thorough tests of my system and a written report on the test results to me in order to facilitate*

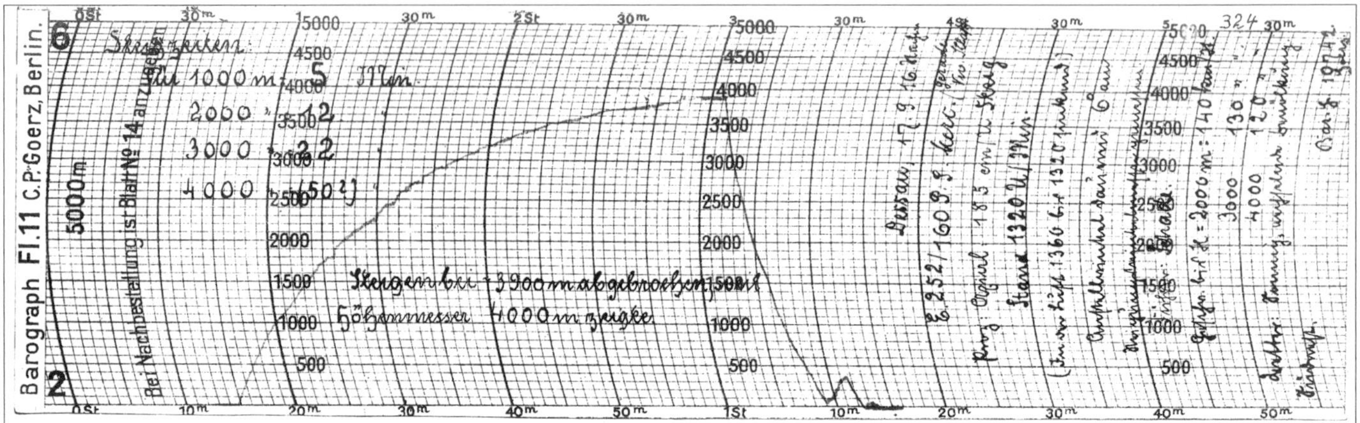

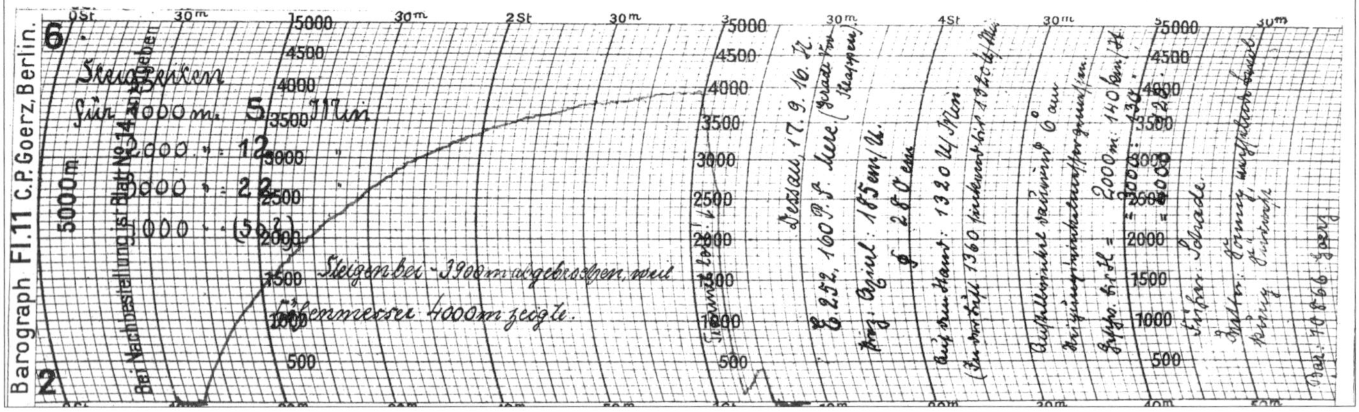

Above: Barograph recordings for 17 September 1916. The speed was added by hand. (Junkers Album via R Zankl)

Facing Page, Top: Taken in November 1916, this photograph shows E.251/16 and E.253/16 with the enlarged rudder, in the factory with another dark painted monoplane behind with the pale painted fuselage of E.250/16 lying next to it. Note the wing and tail structure suspended from the ceiling rafters. Compare with other factory photographs.

Below: The dark camouflage scheme would seem to identify this J 2 as either E.252/16 or E.253/16, however E.251/16 also appeared in such a colour scheme to confuse the issue. The lower cross insignias do not have a white background. (Junkers Album via R Zankl)

424
E254/16
E253/16

FREITRAGENDER
TIEFDECKER J2
1915

Left & Preceding Page, Bottom: A J 2 ended up in the Junkers Technical museum painted in a non-representative colour scheme.

Above: A page from *Die Junkers Lehrschau* showing the J 1 and J 2.

attracting private capital for the enlargement of my company.

4 Hiring more engineers and workers.

In consideration of the described conditions, I would be greatly thankful for a speedy decision.

Signed H Junkers

Junkers placed the following "Proposal for additional support" as an enclosure with his letter:

If formal apprehensions were to prevent further support, then recognition of the importance of the matter (the production of all-metal aircraft) *should allow even substantial difficulties to be removed, like in many similar cases in this war.*

The request for support of my activities should be even more justified as it is not merely the question of the development of a certain aeroplane system but also of the continued existence of the cited Forschungsinstitut *from which the aeroplane descended and which I, like before as expressed in my letter dated 27 October 1916, put at the service of the interests of our country.*

The great value lately attached to the systematic scientific research, also shown by the fact that even during the war substantial funds have been spent on planned and begun institutes, lets me hope that a research facility serving similar ends will not be destroyed as it is already existing, fully equipped

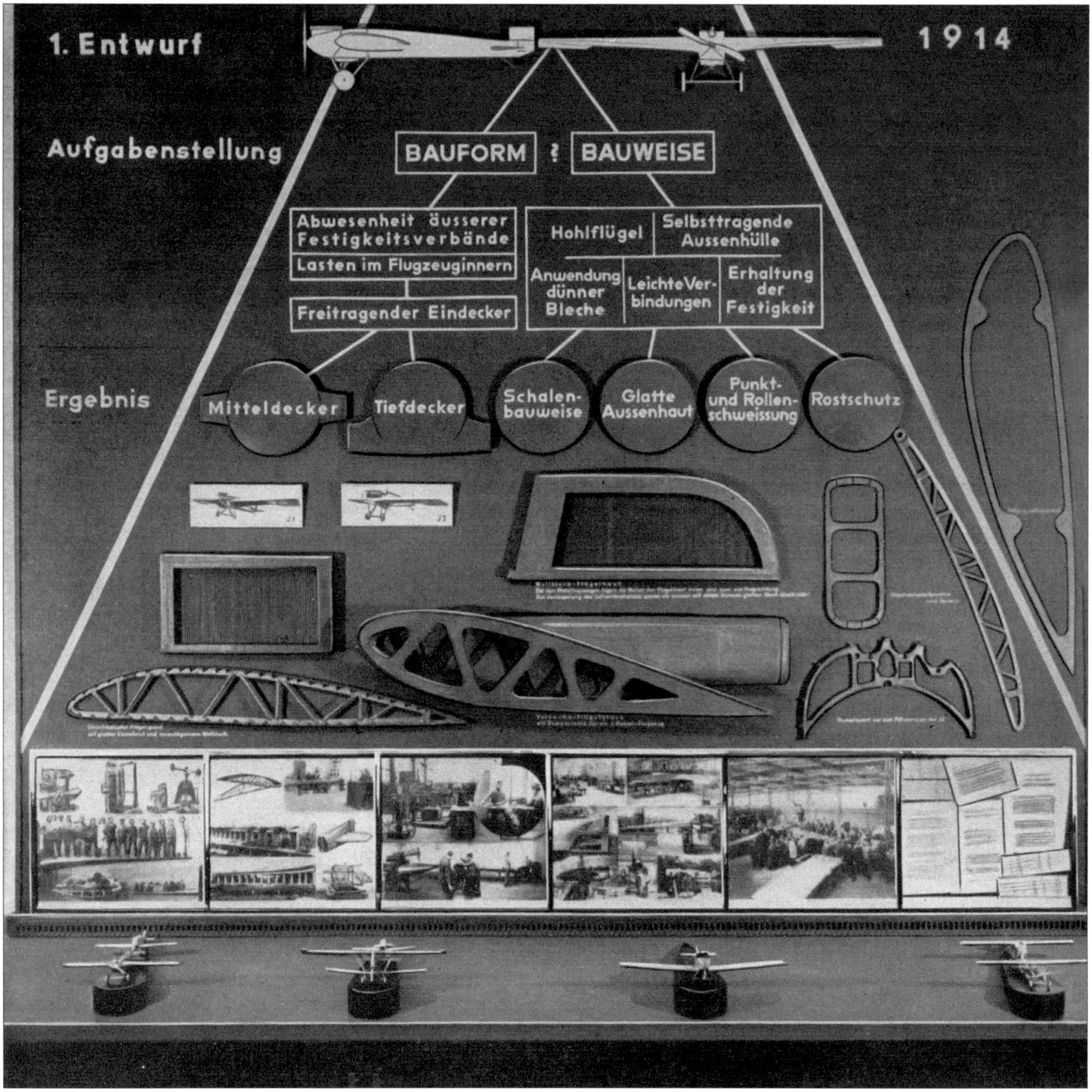

Above: The display in the Junkers Technical museum between the wars showing the development of the J 1 and J 2.

and operating, has not cost the State anything and has already shown its vitality.

Therefore, I request assistance in the form of these necessary funds.

Considering the favourable results and the aims being for the use of the military, it does not appear to me unreasonable that (the authorities), like in many other similar cases in generous ways, would recompense at least part of the considerable costs that I have borne for the Forschungsinstitut *and the completion of the aeroplane* [assuming he is referring to the J 1 here] *and for those which I would need to develop in continuing this work in the interest of the Army.*

In order to maximize the promotion of the construction of my aeroplanes and especially of those areas, which so far have been neglected, where

Above: The Junkers J 4 prototype with two E types in the background.

the particular characteristics of my system (of construction) holds out the prospect of still greater advantages for giant aeroplanes, seaplanes and flying boats, etc. It is also desirable to attract other organizations to utilize our existing manufacturing facilities.

The success of these efforts naturally to a great extent depends on the professional evaluation of my aeroplane manufacturing system by the military authorities who care, not only about the immediate usefulness but also the principal importance of the system and its possible future development. Without effective support from the military authorities the danger cannot be warded off that my entire company, with the aeroplane factory, will fall into in strange hands, whereby the continued development of the aeroplane and the independence of the Forschungsinstitut *will be obstructed, making the company a slave under profiteering.*

He titled his next enclosure "Difficulties in spite of given support".

It is deplorable that we only have a small staff of engineers, assistants and workmen available, otherwise the ordered six aeroplanes had been completed long ago but also had the development of aeroplanes progressed with the climb rate demanded today. But with the small group of around 100 workers, which are mostly occupied with training new methods, modifications etc, it is not to be expected to have an efficient aeroplane production. In these circumstances, it is necessary to abstain from design and workshop work with large aeroplanes. In this area (the construction of large aeroplanes) the particular advantages of the construction techniques manifest themselves to a higher degree and justify the definite expectation that the solution of a problem will be found in new ways and quickly resolved.

My energies and those of my colleagues has been taken up with the difficult task of recruiting personnel, workers, material, funds etc., leaving no time to work on the promising areas of research that the system of manufacture leads to.

It is urgent that these difficulties be alleviated in order to free my staff for tackling the important technical issues.

The difficulties of creating an efficient mass production line of metal fabricated aeroplanes caused by lack of workers and assistants has squeezed my funds and credit so far that I cannot continue on with the research at my own expense to build the aircraft which are of interest to the Army.

To my earlier application for credit for the enormous expenditures relating to the construction of the aircraft for the Army, I received the answer that this could not be done until after the tests of the new type. This condition is now filled. The first prototypes have proven both the doubt the strength and the required qualities in flight, and from the armament point of view. Even if the tests at the Front still have not taken place, the already available results should make it possible for the Army authorities to estimate the great significance and the practical usefulness of my system and justify more support rather than the loss of time through the further "wait-and-see time" (of the prototypes before ordering series production). This policy of the Army causes greater harm to the delivery of a greater number of aeroplanes of this system than any harm caused by the failure of the prototypes at the Front.

The support so far promised and also given by means of orders, which should have presented the suitable road to redressing the difficulties, has to my great distress failed as the production to a greater extent could not be achieved due to the difficulties with personnel, as previously described, and lack of suitable workers.

Even if my past and present extraordinary endeavours, production with praiseworthy support from the War Ministry, Idflieg and other Army authorities should be expanded for a hopefully considerable improvement of these conditions, this could not be expected in the foreseeable future.

Granting further orders is a desired assistance and an indispensable necessity in order to maintain the present manufacturing facility, that has been created with great difficulty and at high cost, to develop large-scale production and design.

In order to continue all the earlier mentioned extensive systematic activities to advance the technological of all-metal aeroplanes, there is a need for considerably greater financial resources than I, my entire company and all other assisting sources can raise.

As a further enclosure Junkers included the following history of his work.

In February 1915 I placed myself, my assistants and means, in particular my aerodynamic research facility, totally at the disposal of the War Ministry.

The War Ministry tasked a commission to inspect the works at Dessau and the plant at Aachen, and make a report of what was seen. As a result of this report I received an order for the first test prototype. The main purpose of this order was to make it possible for me to obtain material, workers and support for the continued development.

Thorough work in the aerodynamic institute led to the design and then the creation of this test aeroplane in my own workshops at Dessau. During the testing at Döberitz at the beginning of this year, this first aeroplane showed excellent flight characteristics and confirmed brilliantly the correctness of the guide-lines for the design based on the through preliminary work. These results of the first test flight are all the more noteworthy as they were obtained not only according to quite new points of view, using new materials with less than optimal characteristics and dimensions which were difficult to procure and to work with, but also without professional experience in the manufacturing techniques (that had to be developed)*, and under the most difficult personnel and working conditions.*

In October 1915, I addressed a demand to the War Ministry asking for a thorough review of my metal aeroplane and for (official) *support based on that review. I was obligingly promised that this wish would be considered and later, after completion of the prototype, got support corresponding to the prevailing conditions by being awarded the first order which was for six aeroplanes. The first of these aeroplanes have meanwhile been completed.*

Of these six aeroplanes, the first was successfully subjected to static testing at Adlershof; the safety factor being 4,95 against 4,5 as prescribed when ordered. The flight tests by Ltn *Mallinckrodt confirmed the correctness of the high speed and the surprisingly good general flight characteristics shown by the first prototype. Concerning the climb rate, which was not of decisive importance at the time the specifications were presented, the requirements had been extraordinarily revised: the climb time was reduced by half. Therefore, it is not surprising that the second prototype in this regard is unable to meet the new requirements. This is even less unexpected as the novelty of the material, design, method of work etc., naturally caused difficulties resulting in the comparatively high weight of the first prototypes. These circumstances* (the high weight) *had, obviously, not been considered, the notion being that the great weight* (of metal construction) *is an unavoidable disadvantage which cannot be removed.*

During a talk with Hptm *Wagenführ*[4] *on 11 August this year at Idflieg, I explained in detail the technical reasons why the climb rate of the second aeroplane still had not reached the desired figure. I confronted the ideas that this remaining deficiency should prove detrimental the manufacture of all-metal monoplanes. I demonstrated, by means of a great number of graphs from model tests in wind-tunnel, that aeroplanes build to the aerodynamics of my system can be expected to show important advances on all other existing aeroplanes with reference not only to speed and rate of climb but also regarding safety and manoeuvrability.*

Obstructions, factually without proof, have been placed in our path and we implore the Ministry's support. The proof of the correctness of my assertion would already be a great number of such aeroplanes if not for the obstructions placed in our way.[5]

This request has been presented in detail as it gives an idea of the circumstances of the German aeronautical industry at this early period. Tradesmen were in short supply and as were technicians and engineers. Materials do not seem to have been a problem if the official requisitions were available.

Junkers himself was passionate in pursuing his goal and this comes across in his writing. He could see the problem that *Idflieg* seems to been blind to – metal aircraft production took a long time to set up as a production line, unlike wooden aircraft, and producing a couple of prototypes, waiting for them to pass all their tests and then to have to wait until they had been tried at the Front meant the loss of time, the loss of workers and the dismantling of the production line.

Endnotes Chapter 3

1. The E stood for *Eindecker* (monoplane). Although referred to as the E.I or E.II in some references, at that time there was no such designation, aircraft were identified as Junkers E.250/16, etc.
2. The *ZAK* Report gives a speed of 205 km/h.
3. Green and Swanborough give the date as the 23rd.
4. Felix Wagenführ was the commander of *Idflieg's* testing department.
5. The problems with skilled personnel were to become worse as time progressed. In the build-up for the *Amerika Programm* the aircraft industry did not receive the 2,000 workers it was promised, and those that they did receive were often poor quality. After the strikes of January 1918, AEG and LVG were placed under military direction and many workers were inducted to the military. Morrow Jr, JH. *German Air Power in World War I,* University of Nebraska Press, USA, 1982. PP102–103.

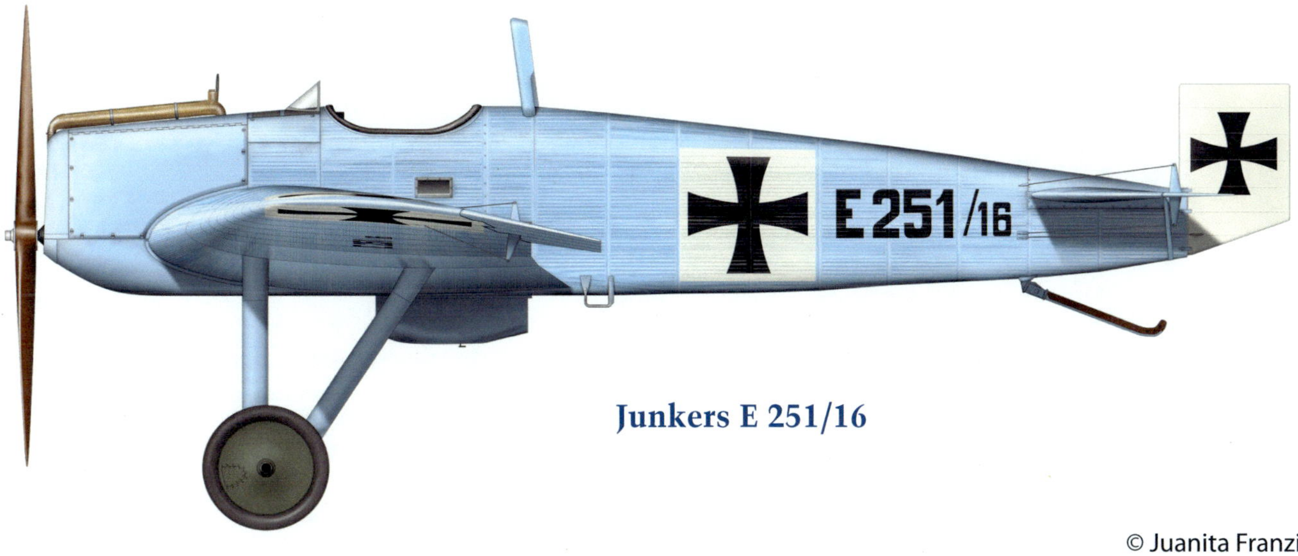

Junkers E 251/16

Above & Below: *Aus der deutschen Luftbeute: Das "Ausschlachten" der erbuteten Flugzeuge, von denen die noch gebrauchsfähigen Teile verwendet warden.* (From the German air booty: The salvaging of wrecked aircraft for still usable parts.) In the foreground is the fuselage of Armstrong Whitworth F.K.8 B263, "Punjab 22 Simla Hills"of No. 10 Squadron, RFC, captured on 24 November 1917. Bristol Fighter and Sopwith Camels can be identified amongst the piles of aircraft parts. The close-up shows a Junkers J 2 without motor dumped as well. The serial appears to be 252/16. The location is thought to be Johannisthal aerodrome. How the J 2 came to be abandoned here is not known.

Junkers E 252/16

Left: Theo Osterkamp.

Right: Gotthard Sachsenberg.

4. Junkers J 3 – The Search for a New Form of Construction

Above: The J 3 was completed to this stage before it was cancelled. The J 3 would have been the first aircraft constructed of duralumin and featured the corrugated skin. Note the J 2 E type wings and fuselage against the wall to the righthand side, possibly from E.251/16. Displayed in the Junkers Museum post-war the J 3 airframe was destroyed in World War II. (SDTB)

Work on the J 3 was ongoing while the J 2 was entering its test phase, but was suspended after the fatal crash of E252/16. The unfinished airframe survived until World War II. The J 3 was built with duralumin and the outer skin was to have been the first to be fitted with the famous corrugated load bearing panels over the internal tubular framework. It was realised that while the construction of the wings of the J 1 and J 2 were very strong, the method of construction meant that they were very heavy. The weight proportion of duralumin to sheet steel was 1:3. To meet the high bending moments of the wing Junkers adopted a trestle of tube framework as the internal structure for the wings with the corrugated surface taking only the local and part of the torsional stresses. The aircraft was to have been powered by a 160-hp Oberursel U.III rotary engine. The fuselage and wings were completed by November 1916 with the wings bearing the corrugated duralumin panels.

Junkers had to develop new methods of fabrication as the duralumin panels could not be welded as it reduced the strength of the material. The skin was corrugated to increase its strength and while this increased drag it was not significant given the speed of contemporary aircraft. The wings were formed around a number of tubular spars that decreased in size towards the tips where the loads were less. These were not spars in the traditional sense, and were joined by telescoping and splicing, with a girder constructed between them giving a light, strong structure with the corrugated skin taking shear forces. Junkers built experimental wing sections that were tested enabling him to evolve a reliable structure.

The cost of working in completely unknown areas, working out new methods of fabrication, and developing new metallurgical processes was very expensive. The J 3 was proposed as two versions. The first was a single seat machine of span 11.50 m and a length of 6.70 m. The second was a two-seat machine of the same general outline with a span of 13.00 m and a length of 7.70 m. Power was to be the 160-hp Oberursel twin row rotary engine in both cases.

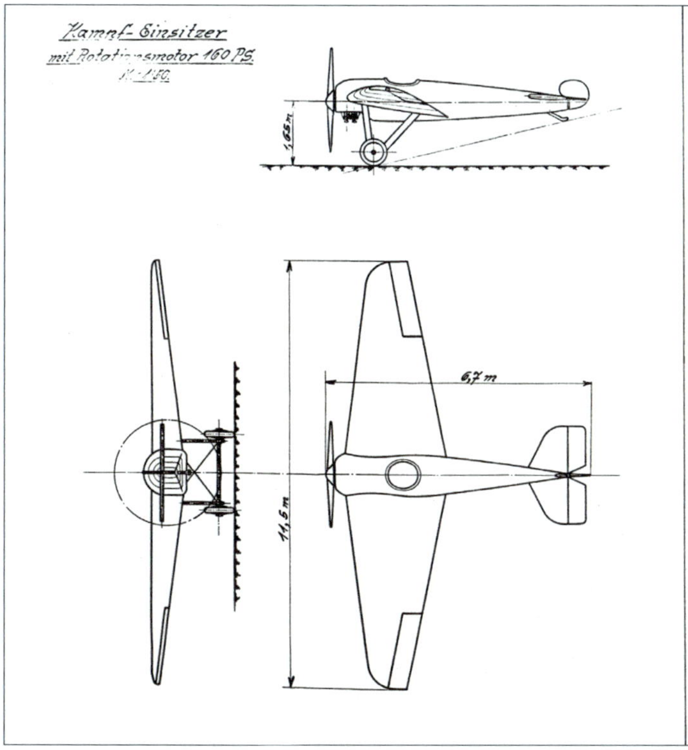

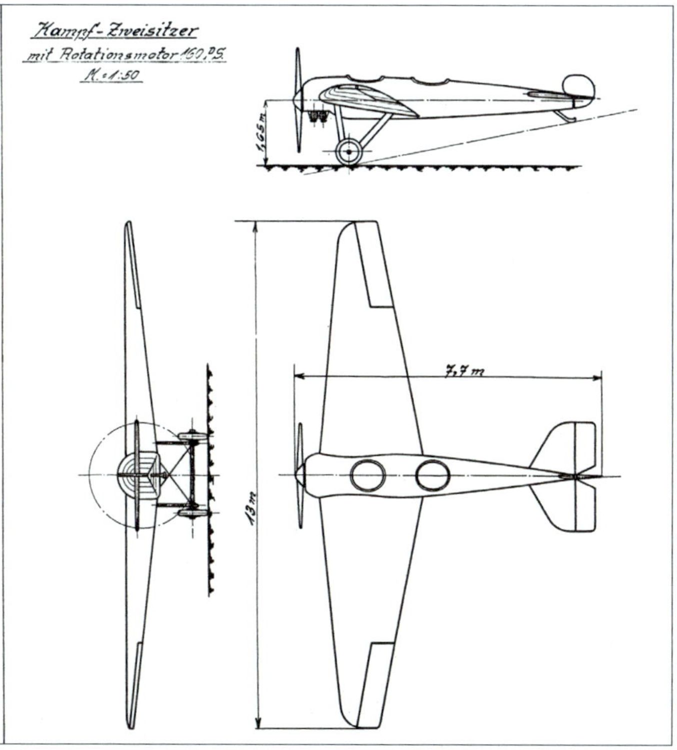

Above Left: Junkers' drawing showing the proposed single-seat J 3 monoplane.
Above Right: Junkers' drawing of the proposed two-seat version of the J 3 showing the same general plan as the single-seat version.

The project was abandoned when *Idflieg* would not provide financial support for the proposed aircraft. However, Junkers did receive an order for an armoured ground attack aircraft, and on 26 December 1916, he received an order for three monoplane fighters. The order for these prototypes did not specify whether they would be single-seat or two-seat fighters and the way was open for Junkers to continue with his work on the monoplane.

The partially completed J 3 airframe was initially displayed after the war in Junker's exhibition hall at his Dessau plant.

Specifications J 3		
Type	**Single-Seater**	**Two-seater**
Dimensions in m		
Span	11.50	13.30
Length	6.70	7.70
Engine	160-hp Oberursel	160-hp Oberursel
Source: Junkers' drawing 22.09.1916.		

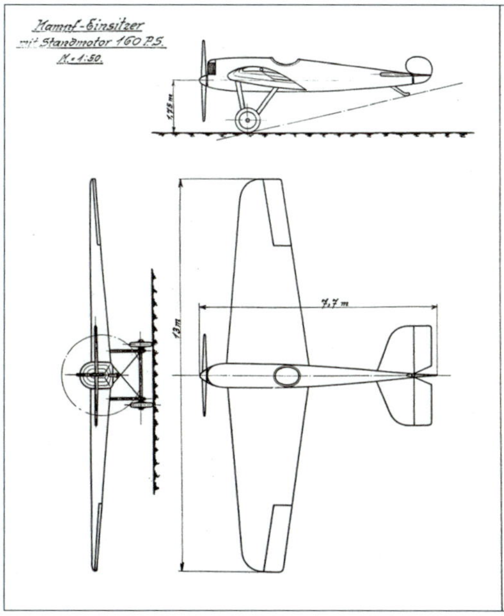

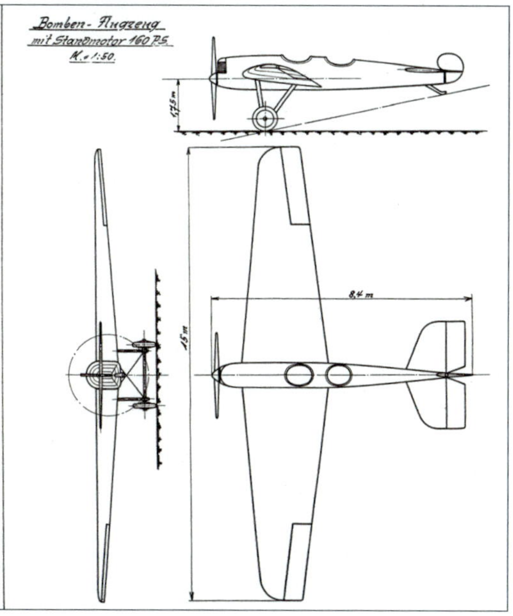

Right: Junkers also considered using a 160 hp water-cooled engine for the aircraft. That would have been more reliable but heavier. This drawing shows the proposed straight-six powered-single seat J 3 monoplane.

Far Right: Junkers' drawing of the proposed two-seat version of the J 3 showing the same general plan as the single-seater with water-cooled engine.

Page from *Die Junkers Lehrfchau* showing the J 3 on display in the Junkers technical museum post-war. The other photographs show the way the corrugated metal was developed and tested.

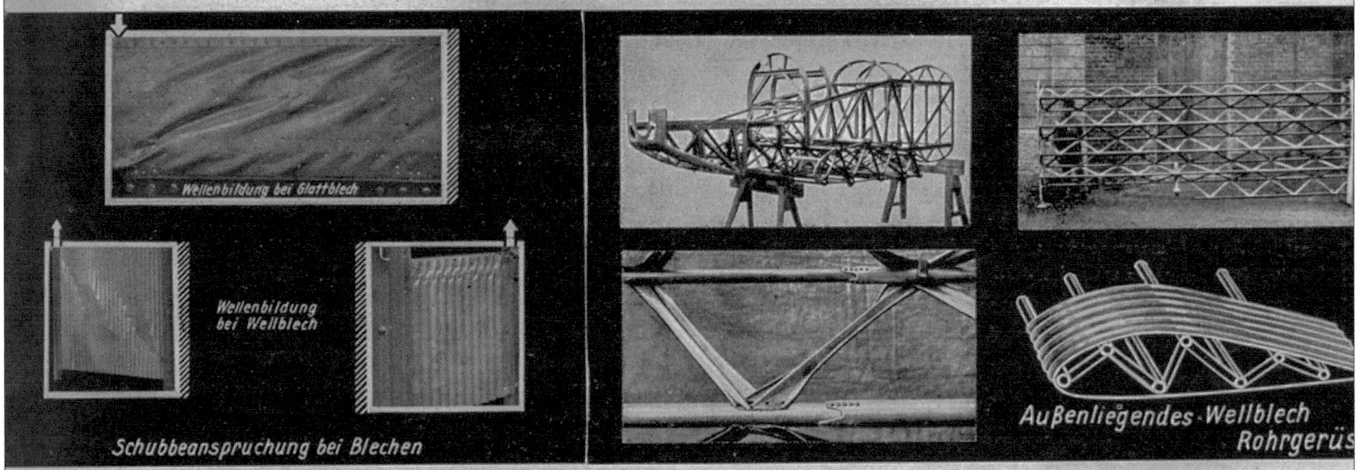

Schubbeanspruchung bei Blechen.

Flügel- und Rumpfbauweise.
Erstes Junkers-Duralumin-Flugzeug J 4.

5. Junkers J 4 (Military J.I) – First Serial Production of a Metal Aeroplane

Above: The prototype J 4 on 5 January 1917. The struts have not all been given their fairings. As the armoured steel panels were not available, plywood was substituted to enable flight testing. (Junkers Album via R Zankl)

The war in France had developed into a stalemate by 1915. The vast armies were dug in for miles of trenches facing each other over "No Man's Land." Each side hoped for the breakthrough that would set loose the cavalry and start a war of movement again. Information was gained by reconnaissance aircraft but in the mass attacks of infantry the wooden and fabric aircraft of the day could not long survive flying low over the battlefield trying to identify the location of troop positions, both their own and those of the enemy. According to the late Peter M. Grosz the Germans were impressed with the results of British Contact Patrols during the Battle of the Somme from 24 June to 26 November 1916. As a result, the Germans developed a requirement for an aircraft to perform *Infanteriefliegerdienst* duties that required very low flying over the attacking troops.[1] What was needed was an armoured aircraft that could survive in this maelstrom. This was the J Type, an armoured two-seater.[2]

Idflieg issued specifications in late 1916 for such an *Infanterie-Maschinen.* It was to be a heavy armoured biplane. Albatros and A.E.G. proposed armoured versions of their successful C.XII and C.IV types respectively.[3] Junkers also received the specification. The Junkers proposal, the J 4, was an entirely new aeroplane. Junkers placed the design into the hands of *Dr-Ing* Otto Mader[4] with the team of Reuter[5], Steudel[6] and Franz Brandenburg[7]. To this must be mentioned the name of Madelung who was *Idflieg's* representative at Dessau. Junkers apparent lack of interest in the machine was probably due to *Idflieg's* insistence that the machine be a biplane whereas Junkers had proposed a monoplane More correctly, it was a sesquiplane, with the lower wing of much shorter span and chord than the top plane.[8]

Above: J 4 prototype in the snow on Döberitz airfield, 28 January 1917. This machine did not have the armour plate installed; plywood panels were used instead to enable the flight testing to go ahead.

Above: The prototype J 4 that performed the Aldershof flight evaluation on 7 May 1917. The fin and rudder have been enlarged after the first flights. Note the instrument pitot on the lower starboard wing tip. A single V-strut braces the fin. The corrugated wings and tailplane appear to be natural metal or to have been given a clear coat and the rudder is covered by canvas. The fuselage appears to be the one colour over metal and fabric panels. Despite being forced to adopt a biplane layout, Junkers used cantilevered wings. (SDTB)

Above & Below: Additional views of the prototype J 4 that performed the Aldershof flight evaluation on 7 May 1917. T (SDTB)

Above: A view of the prototype J 4 in the factory; it does not have struts to brace the fin and does not yet feature a balanced rudder.

Above & Facing Page, Top: Prototype J 4 with early fin and rudder. The gunner's cockpit was located under the trailing edge of the upper wing and he could not operate his gun correctly.

Below: Possibly an early production aircraft this J.I has the triangular fin with one support strut, a balanced rudder, and enlarged semicircular cutout of the center-section trailing edge to give the crew more clearance. The gentleman posing with the machine is the engineer who helped design it, *Diplom-Ingenieur* Franz Brandenburg. 3 May 1917.

Above: Probably the same aircraft as the photograph was taken the same day, 3 May 1917. The fields for the wing crosses do not encroach onto the ailerons. (Junkers Album via R Zankl)

Above & Facing Page, Top: Junker's photograph No.667 of 23 May 1917, shows the modified rudder and fin. The fin is now larger with two struts bracing it on either side.

Above: The machine now has balanced ailerons as well as the enlarged strut-braced fin and balanced rudder. Lift handles have been provided near the tailplane. Junker's photograph No.699 of 12 June 1917.

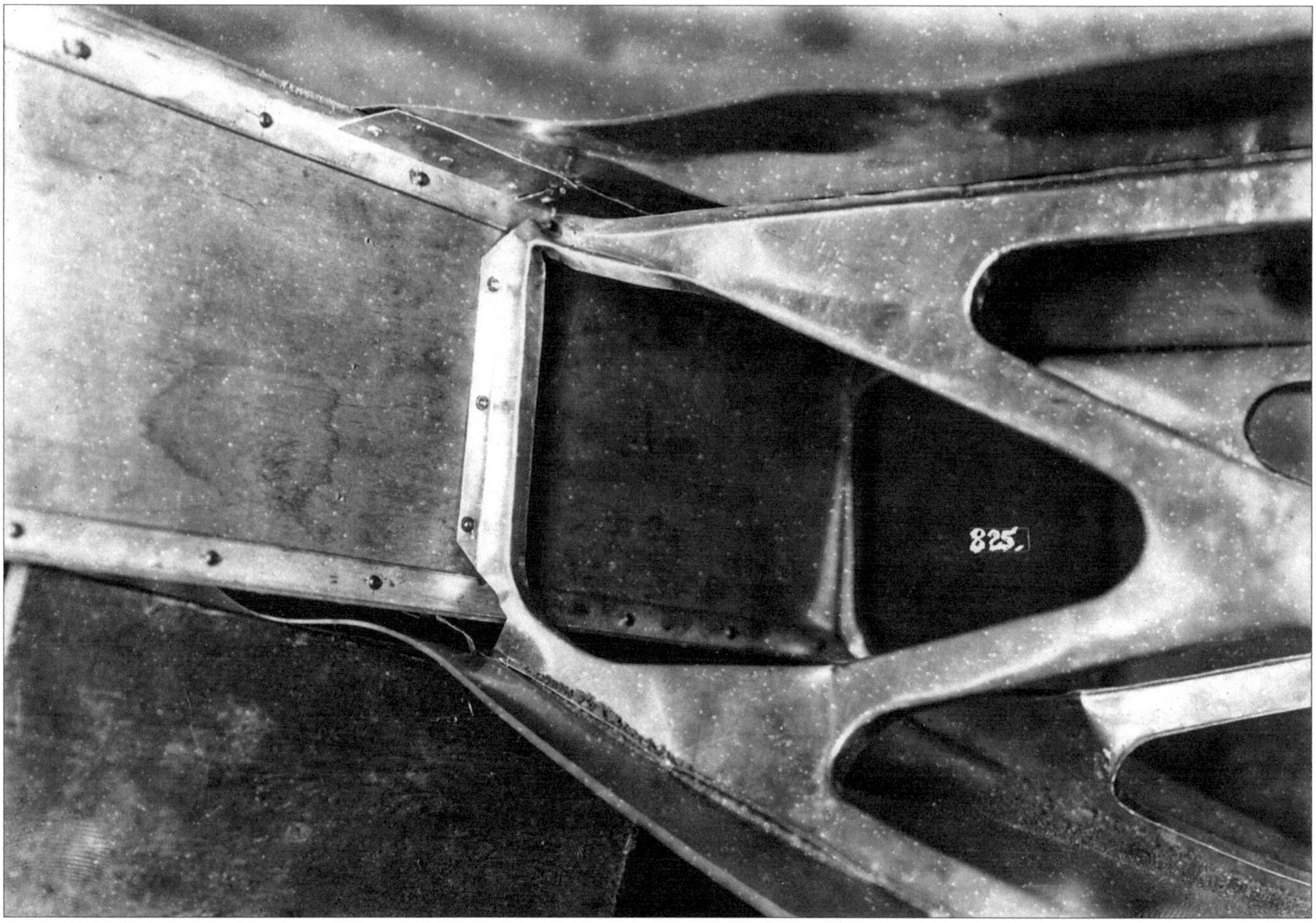
825.

Zuwiderhandlungen gegen die
Unfallverhütungsvorschriften
Zuwiderhandlungen gegen die
Unfallverhütungsvorschriften
48
888.

Above & Facing Page: Fokker photographs of experimental metal wing structures. The ply-covered wooden wing was to be used by Fokker until World War II.

Three prototypes were ordered on 3 November 1916, (J.425/17 to J.427/17), with the first being completed by the following January.[9] This machine was powered by a 200-hp Benz Bz.IV engine and had plain ailerons. Of sesquiplane layout the machine was first flown on 28 January by *Flugzeugführer* Arved von Schmidt at Döberitz, the machine having been shipped to Adlershof at Berlin for type testing. At this time, the armour plate was not installed, plywood sheet being substituted instead. It performed so well that 100 were ordered on 19 February 1917, however it was to be May 1917 before the armour plate met specifications and was available for attaching to the aircraft.

Junkers had begun to dismantle the production line by the time that the J 4 proved itself at the Front. Now Junkers had an order but production output was lost. *Idflieg* did not understand that all-metal aircraft required special preparation to achieve serial production. Materials had to be stockpiled, qualified manpower and financial backing obtained. This was to come to a head later in the J 9 (D.I) production order.

The first construction batch of the J.I biplanes was given the serials J.100/17 to J.149/17 and was to be powered by the Benz Bz.IV straight six engine. The machine passed its type test at Adlershof on 7 May. This may have been the prototype with armour installed. The prototype was allowed to go to the front for evaluation. Another of the prototypes was the subject of load testing that saw the rear fuselage being strengthened. Flight testing continued with a prototype at Döberitz, the lessons learned undoubtedly being incorporated into the machines being assembled at the Junkers factory. Amongst changes to the production machines was a redesigned vertical fin and overhung balanced ailerons, and a balanced rudder. The fin and rudder were fabric covered for the initial batches.

The first production J.I (100/17) was at the front by August 1917, as the first operational flight was probably that of *Flieger-Abteilung* 19 on 1 August. The ability of the J.I to take punishment that was easily repaired with the special tool kit that came with each aircraft, its better speed over the Albatros and A.E.G. J.I biplanes, made it a welcome

Above, Below, & Facing Page: The first production J.I, J.100/17, was rolled out in a simple one-colour upper surfaces scheme with an unusual presentation of the serial and the Junkers logo on the wheel covers:- "Junkers & Co Dessau." A single strut braced the fin with like the rudder was canvas covered. The louvres on the nose changed as production proceeded. Note the flare pistol and cartridges in the rack on the port side of the fuselage near the gunner's cockpit. (AHT AL0466-054, 056 and 057).

J.100
17

Above: J.101/17 at *Flt Abt (A)* 250 with the serial more in keeping with standard practice. The J.I was used operationally during September and October 1917 for operational evaluation, first at *Fl Abt* 19, then *FAA* 250.

Below: J.101/17 in the field. Note the lower struts are a light (under surface) colour.

Above: J.112/17 has had the curved cross changed to the intermediate type by the simple method of squaring up the curved fuselage cross. The wing crosses, on the top surface at least, are still the curved variety. The canvas covers on the ground are for covering the engine and cockpits as on the J.I in the background. (AHT AL0613-017)

Below: J.113/17 displays the tailplane connection in this view of the aircraft after turning over. The ground looks solid but the heavy Junkers could find any soft spots with this as a likely result. Interesting is a light spot under the missing stabilizer showing that the second (grayish-violet) color was applied later. (AHT AL0014-07)

Above: The ground crew appear to be holding J.802/17 back as the engine runs up in this propaganda photograph. The machine has full serial application. The crosses are the interim style and carried under the lower wing. Note the streamer attached to the wing. The "8" on the fin is probably the aircraft's position in the unit. The fuselage band appears to be white and is a squadron marking. The machine appears to have a single solid colour overall upper surfaces.

addition to the front-line squadrons. While its manoeuvrability was not good, it was satisfactory.

The J 4 was the first Junkers aircraft to be built in quantity. While the early Junkers aircraft were all picked up by train and moved to Döberitz near Premnitz for test flights, the production order for the J 4 led to the construction of an airfield of 1,000 by 380 metres at the Mosigkauer Heide between Mosigkau and Dessau-Alten. Two hangars were constructed and this facility was used up to the 1920s when larger aircraft needed more room and a new airfield had to be found.

The machine soon proved itself and a further 100 were ordered in December (J.800/17 to J.899/17), 50 more by July 1918, and another 100 in October. Linke-Hofmann company was given a contract for

J 4 (Military J.I) Orders and Serials

Date of Order	No.	Known Serials
3 November 1916	3	Prototypes. J.425/17 to J.427.
19 February 1917	50	J.100/17 to J.149/17
December 1917	100	J.800/17 to J.899/17
20 June 1918	40	J.576/18 to J.615/18
12 July 1918	10	J.716/18 to J.725/18
October 1918	100	*Bestellunmmern* J.726/18 to J.805/18 only were assigned. The last 20 were cancelled and CL.1 monoplanes ordered instead.

Facing Page: Another photograph of J.802/17 showing the loading of supplies that will be dropped to front line troops. In practice the supplies were contained in wicker baskets to prevent damage when dropped. Note how the light under-surface colour ends at the metal forward armoured shell.

802/17

Junkers J.I Deliveries Ended June 1919

Month	J.100–149/17		J.425–427/17		J.800–899/17	
1917						
Aug	100, 101	2				
Sep						
Oct	103, 105	2				
Nov	102, 106–108	4				
Dec	109–112	4				
1918						
Jan	113–116, 119–122	8				
Feb	118, 123–126, 129–131, 133–136	12				
Mar	127, 128, 132, 137, 138, 140–147	13				
Apr	139, 149	2			800, 801, 808	3
May	148	1			802, 805, 806, 809–811, 813–815, 818, 824, 826, 838	13
Jun					817, 819, 821–823, 827–828, 832–836,840, 843, 850, 858, 861	17
Jul					803, 804, 807, 820, 829, 830, 837, 839, 848, 853, 854, 857, 859, 860, 862–864, 866–867, 869, 874	21
Aug					812, 831, 841–842, 844–846, 851, 855–856, 865, 870–871, 873, 875–877, 882–884, 889–890	22
Sep					816, 825, 847, 878–881, 896–899	11
Oct					852, 872, 891, 893	4
Nov					885–887, 892	4
Dec						
1919						
Jan						
Feb						
Mar						
Apr						
May						
June						
Not known	104[1], 117[2]	2	425[3]-427	3		
At factory						
In storage Jonitz					888, 894-895	3
In storage Tuchmann						
Cancelled						

Notes: 1. J.104/17. No shipping date recorded. 2. J.117/17 reported in the Field. 3.J.425/17 was load test specimen, no shipping date recorded. 4. J.749/18 had wings of increased span and corrugated duralumin

Junkers J.I Deliveries Ended June 1919

J.576–615/18		J.716–725/18		J.726–825/18		Mthly Total
						2
						0
						2
						4
						4
						8
						12
						13
						5
						14
						17
						21
						22
576–579, 583–586, 593–596	12					23
597–599	3					7
605	1					5
						0
				753, 755	2	2
						0
				747, 748, 750–752, 754, 758, 760, 762, 763	10	10
591, 592	2					2
		725	1	726, 729, 730, 741, 744	6	7
608	1	724	1		2	2
						5
				749[4]	1	1
580–582, 587–588, 602–604, 612	9	717–723	7			19
589–590, 600, 601, 606, 607, 609	7			727, 728, 731, 735–740, 742, 743, 745	12	19
610, 611, 613–615	5	716	1	732–734, 746, 756, 757, 759, 761, 764–805	51	57

covered rear fuselage. A. The order for J.806/17–825/17 was changed to 20 CL.I monoplanes. B. J.849/17 crashed on 24 October 1918. C. J.868/17 No shipping date recorded. [Notes A, B, C not referenced in table.]

Above: J.833/17 has the upper surfaces in a two-colour scheme with late narrow crosses. Note the white outlined "7" on the fin. The machine was accepted at Dessau in the week of 10 June 1918. A wing with the thick intermediate national insignia resides in the hangar in the background.

Above: J.102/17 on a typical operational field.

Above: The result of a lost tyre – J.102/17 on its back. The starboard lower wingtip has suffered damage when the tyre gave way. Struts appear to be the same light colour as the lower surfaces but these early machines had green struts, and this effect is apparently due to the lighting conditions. The photograph is dated 10 September 1917. It was sent back to the Front on 28 January 1918. Very light blue underside paint at wings and rear of the fuselage was added during prior repairs. Underwing crosses of the upper wings were over-painted. (SDAM)

Right: 104/17 appears to have buried its nose well into this field. It did not take much of a soft spot to cause a problem for the Junkers two-seater. Note the ubiquitous wooden ladder. Ladders and wooden horses were as plentiful as dogs around airfields for this type of work and general maintenance.

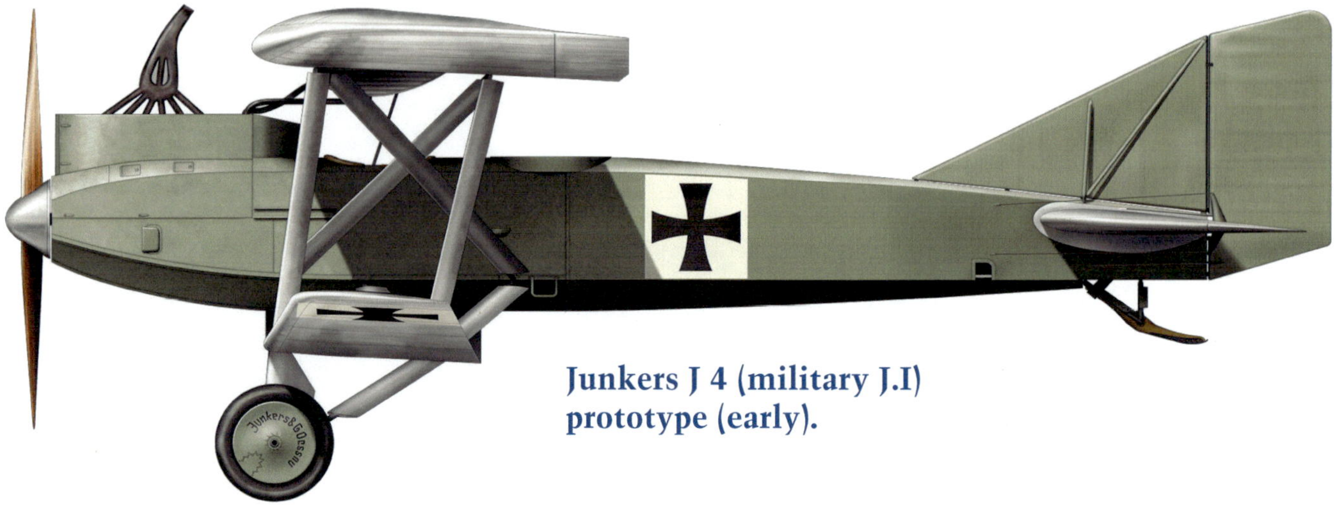

Junkers J 4 (military J.I) prototype (early).

Junkers J 4 (military J.I) prototype (late).

Junkers J.I J.100/17

Junkers J.I J 101/17

Junkers J.I J119/17

Junkers J.I J 134/17
Lt. Wilhelm Schreiber/Fwbl. Ernst Schäfer
Flieger-Abteilung (A) 221
May 1918

Above: Working on J.138/17 in the field. This machine had left the factory in the week of 5 March 1918, and was attached to *Fl Abt* 17. On this machine the interplane struts appear to be a dark colour. The application of the cross on the bottom of both wings was not uncommon, being requested by crews as they were fired on by both sides.

Below: The corrugated metal gap cover over the aileron gap was free and relied on the airflow to force the flap down to cover the gap when the machine was in flight. It stands up well in this photograph of J.118/17 of *FAA* 208. The rudder appears to be overpainted a dark colour.

Above: Officers and ground crew pose with J.148/17. Despite the warnings contained in the manual for the type, these men are obviously confident of the strength of their Junkers J.I. The machine is in the two colour scheme. Note the stencils on the nose at the catches that allow the clam shell engine cowling doors to open. Under this it reads *JFa* over *Farb. No 148*, over *[Junk.I]*. Note how the lower surface colour extends onto the upper fuselage. The port lower wing is missing or not fitted. (AHT AL0776-116)

Below: Officers posing with J.817/17. Note the final type of national markings on this early J.I.

Above: Judging by the mounds of harvested material on this field, this Junkers J.I was lucky to come down in one piece. Note the gap between the lower port wing and the centre-section and the opening in the fuselage behind the gunner's cockpit. This machine is in the single colour camouflage scheme. In each case where it appears to be the two-colour scheme a shadow can be seen to cause the variation in tonal values suggesting a different colour. (via R Zankl)

Below: Ground crew pose in front of J.109/17. The early curved cross is carried on the lower surface of the upper wing while the fuselage has a straight sided intermediate cross. Note the cooling slots in the cylinder armoured turret.

Above: Crew and ground crew pose with J.115/17 from *Fl. Abt. (A)* 217. This machine had been accepted and delivered in the week of 30 December 1917. Note the wing crosses are carried under both upper and lower wings. A "3" is carried under the starboard lower wing in black and on the fuselage in white. There appears to be a light coloured band before the tailplane. Streamers are attached to the lower wing tips.

Above: J.I 119/17 of *Fl Abt (A)* 217. The rudder appears to be a lighter colour than the camouflage but that is probably just the play of the light. Note the "lightning bolt" type band and individual aircraft number "9" on the rear fuselage.

Above: *Ltn* Schreiber, *Vzfw* Schäfer, and *Ritter* Meike with J.134/17 of *FA(A)* 221, Flanders, 30 June 1918. The colours of this J.I are interesting. The armoured nose appears to be a darker colour than the rear fabric-covered section that matches the upper wing surface colour; was this a unit marking? The application of the serial to the fin would suggest that the machine has been recovered and repainted in the field. The crosses appear to have been developed from the earlier curved ones and are carried under both wings. What appears to be an "A" is carried on the fuselage. Note lack of a spinner and the large two-story brick house in the background. Schreiber (on wing at left) was awarded the *Pour le Merite*; his pilot Ernst Schäfer is on the wing with him.

Below: Note the style of serial application on this J.I. "Junk J.I (Jfa) 143/17. The thick interim national markings are applied over the two-colour camouflaged upper surfaces. On this machine the struts appear to be the dark camouflage colour. (SDAM)

Above: On some photographs, it is not possible to determine if the colour scheme is a single colour of the two-colour scheme as shadows give the appearance of the two-colour scheme. On the original of J.140/17 the camouflage on the wings and rear fuselage is definitely two colours. In the wooden hangars in the background is a Rumpler two-seater. (AHT AL0079-63)

Below: Rear view of J.140/17 shows the two-colour camouflage scheme. The interplane struts appear to be both a dark and a light colour on the port side. The lower struts from the fuselage to the centre section appear to be a light colour. Note the aileron gap covers on the upper wing. J.140/17 had been shipped to an *Armee Flugpark* in the week of 11 March 1918. (AHT AL0079-06)

Above: Another view of J.140/17. The rudder cross has a white outline on the camouflaged corrugated surface. The all metal corrugated covered rudder became standard from J.140/17.

Above: J.845/17 was at Aldershof in September 1918 for testing. This official *Idflieg* photograph was taken at that time.

Above: J.149/17 leads this line up of five J.I biplanes in the two-colour camouflage scheme and bear the thick interim cross national markings. The rudder is corrugated metal but the fin appears to be fabric covered. The presence of the civilian indicates that these machines have not left the confines of the factory airfield at Dessau.

Above: Propaganda photograph of a heavily retouched line-up of J 4 biplanes with serials in the 800 series.

100 J.I(Li) biplanes in order to increase production, but none were delivered before the end of the war, and most probably manufacture of the metal biplane was not commenced.[10] It was unfortunate that the J 4 did not lend itself to large scale production with the methods then in practice at the time.

According to the late A.R. Weyl, the now Major Felix Wagenführ, the *Idflieg* officer in charge of aeronautical development, decided that as Professor Junkers was ignorant of aircraft production and flying but had developed a revolutionary aeroplane, and Anthony Fokker was a brilliant pilot who knew how to produce aircraft in quantity, therefore a combination of the two minds would be of great benefit to the German production program that was behind in the America Program. It was hoped that this program would produce more aircraft to counter the United States entry into the war on the Allies

Above: This line up is identified as taken at Junker's Dessau field. The two further J.I biplanes are J.594/18 and J.596/18. All three have the late, narrow cross national markings while J.596/18 has the rear fuselage covered with five-colour printed lozenge fabric. All were sent to the front in September 1918.

Above: In January 1919, J.594/18 and J.596/18 were examined by US forces at Villers la Chevres after the Armistice. This photograph appears to be taken at that site. J.816/17 with the fabric missing from the rear of the starboard side looks the same as the unidentified J.I in the above photograph, leading one to suspect that the previous photograph has been misidentified in the past.

side. In theory *Junkers & Co (Jco)* would produce new aircraft designs and prototypes that would be built in series by *Junkers Fokker Werke AG (Jfa)*.[11]

The specialised construction of the J 4 meant that it could not be sub-contracted to other firms whereas contemporary wooden aircraft could have components sub-contracted to wood working manufacturers as well as other aircraft firms. Whether this was the reason or not, Anthony Fokker was called upon to coordinate the activities of the Junkers works. On 20 October 1917, the *Junkers Fokker Werke AG* (*Jfa*) was formed at Dessau with the purpose of producing the J 4 in meaningful monthly numbers.[12] This was the first dedicated aircraft manufacturing facility built at Dessau.

Fokker recorded in his autobiography that in late 1915,

I was asked by the army to co-operate with

Above: Close-up of the interplane struts on J.586/18. Note what appears to be an ordnance rack between the undercarriage front legs." (AHT AL0144-02)

Above: Fitted with a Mercedes D.IVa engine J.849/17 had revised engine cowling panels that did not require a spinner.

Hugo Junkers, father of the thick wing and all-metal plane, in developing an all-metal single-seater monoplane. Professor Junkers is one of the pioneers in airplane construction. His theories have frequently been in advance of his time, but he has not always been able to adapt his laboratory experiments to practical ends of the moment. His single-seater, tricky and unmanageable, was a highly dangerous flying machine until certain alterations in its design were made under my direction.

Above & Below: J.849/17 in full camouflage and markings, bears the late narrow crosses. It stayed at the Junkers-Fokker Factory for flight testing. (AHT AL0459-218)

Above: A 200-hp Benz Bz.IV engine installed in J.587/18. This machine was stored at Jonitz in December 1918. Its eventual fate is unknown. (SDTB)

Above: J.I 750/18 in the field with a Halberstadt CL.IV (Rol). The Halberstadts provided top cover for the low flying Junkers. Note how the late style fuselage cross has split the serial application on J.750/18. Quoos was in service in Turkey to 1918, and this photograph was probably taken in 1919. (August Quoos album)

Above: Front view of a J.I with the late, narrow cross national insignia.

Above: A J.I amongst the organised chaos of an operational airfield. This Junkers has been identified as J.138/17 of *FAA* 287 at Monceau le Neuf airfield.

Above: This crew is showing the anti-aircraft shell damage caused to their J.I. The fabric has been blown away but the young gunner in his armoured shell is unharmed. Note the telescopic sight on the gunner's machine gun and Morell speed indicator fitted to the wing strut.

Above: All the bullet holes in this J.I have been marked with a circle. The J.I could take a tremendous amount of punishment and still return to base. (Piotr Mrozowski)

Fokker does not record which monoplane he is referring to but it is most probably the J 2. He next records that he invested 3 million marks in the Junkers concern to buy a 50% interest. The paragraph is structured such that this appears to be a 1915 decision and not one brought about by the Army in 1917. "At first Junkers welcomed my co-operation."

I persuaded him to construct many parts of his planes of steel-tubing, and urged him to permit the rudder, elevator, and ailerons to be fabric-covered to facilitate manufacture. He refused to give up his all-metal construction for the sake of wartime expediency. Eventually, he built a ground attack and observation plane which was successfully flown, but I had found it necessary to withdraw from a partnership which cost me 1,500,000 marks and many periods of fruitless argument. Junkers was too theoretical, in my opinion, at a time when practice was so much more important than theory.[13]

There appears to have been no real cooperation between the two men and Fokker went his own way with welded steel fuselages and thick wooden wings, while Junkers continued to develop the all-metal aeroplane.[14] In April 1917, while negotiations were taking place about the merging of Junkers and Fokker, Fokker informed Junkers that his interest was in the thick wing patents that Junkers held.[15]

The J 4 in detail

This description is based upon the Ministry of Munitions "Report on the Junker Armoured Two-Seater Biplane Type J.I", notes on G/3Bde/31[16] and observations of J.586/17 in the Canadian Aviation and Space Museum by John Weatherseed. (Imperial measurements quoted are as given in the original British reports). The Ministry report was based on two salvaged airframes. One was reportedly brought down by the French ground fire while the other was obtained by British troops during an advance near La Vacquerie.[17] Both machines were entirely dismantled

Above: Three J.I biplanes outside the airship hangar at Luxembourg post-war. The closest machine has a dark band around the fuselage. DFW C.V biplanes in background. Note the spotted camouflage on the replaced bottom right wing on the nearest aeroplane.

and greatly damaged.

The Junkers was considered "radically different from the usual type of aeroplane, whether considered from the point of view of design or construction." All the vulnerable parts were as far as possible gathered in the armoured portion of the fuselage. There were no bracing wires, the outer wing panels of both wings being cantilevers. The wings tapered in plan and in thickness, becoming thinner towards the tips where the stresses were less.

The fuselage was constructed in two distinct sections. The forward octagonal shaped armoured portion containing the engine, crew, petrol tank, wireless telegraph, ammunition and part of the controls and petrol tank was built up of 5 mm armoured plating on four longerons of 40 mm aluminium tubing. The fairing was of sheet aluminium. The rear fuselage section was constructed of built up duralumin tubes and covered with laced on fabric. The observer/gunner's seat was built into the rear bulkhead of the armoured nacelle. The rear of the fuselage has a curved backrest for the observer. Any parts outside the armouring were duplicated. The main fuel tank (26 gal) was the pilot's seat. The petrol tank was divided into two parts so that the tank would not drain completely if it was punctured. The oil tank was located just in front of the instrument board (Capacity of 10 gal). The nacelle was only open along the top and was constructed of highly tempered steel 5 mm thick manufactured by the *Panzerwerke* of Dillingen. Quick access was allowed to the engine on all sides by armoured panels hinged at their rear ends that allowed them to swing out for access. The armour was not added to the aircraft but formed part of the structure. Unusually for an all metal aircraft the engine bearers were of solid ash braced to the longerons with 20 mm tubes in a girder design. The engine was totally enclosed, except at the top.

The vertical cowl surrounding the engine cylinders was of armour plate. The "armouring is very through, so that the chances of a bullet finding a vulnerable spot are small." The weight of armour was 470 kg and hence the take-off took longer and climb was also slower than contemporary two-seater aircraft. This also meant that the J.I required a good airfield with a solid surface.

The rear portion of the fuselage comprised four

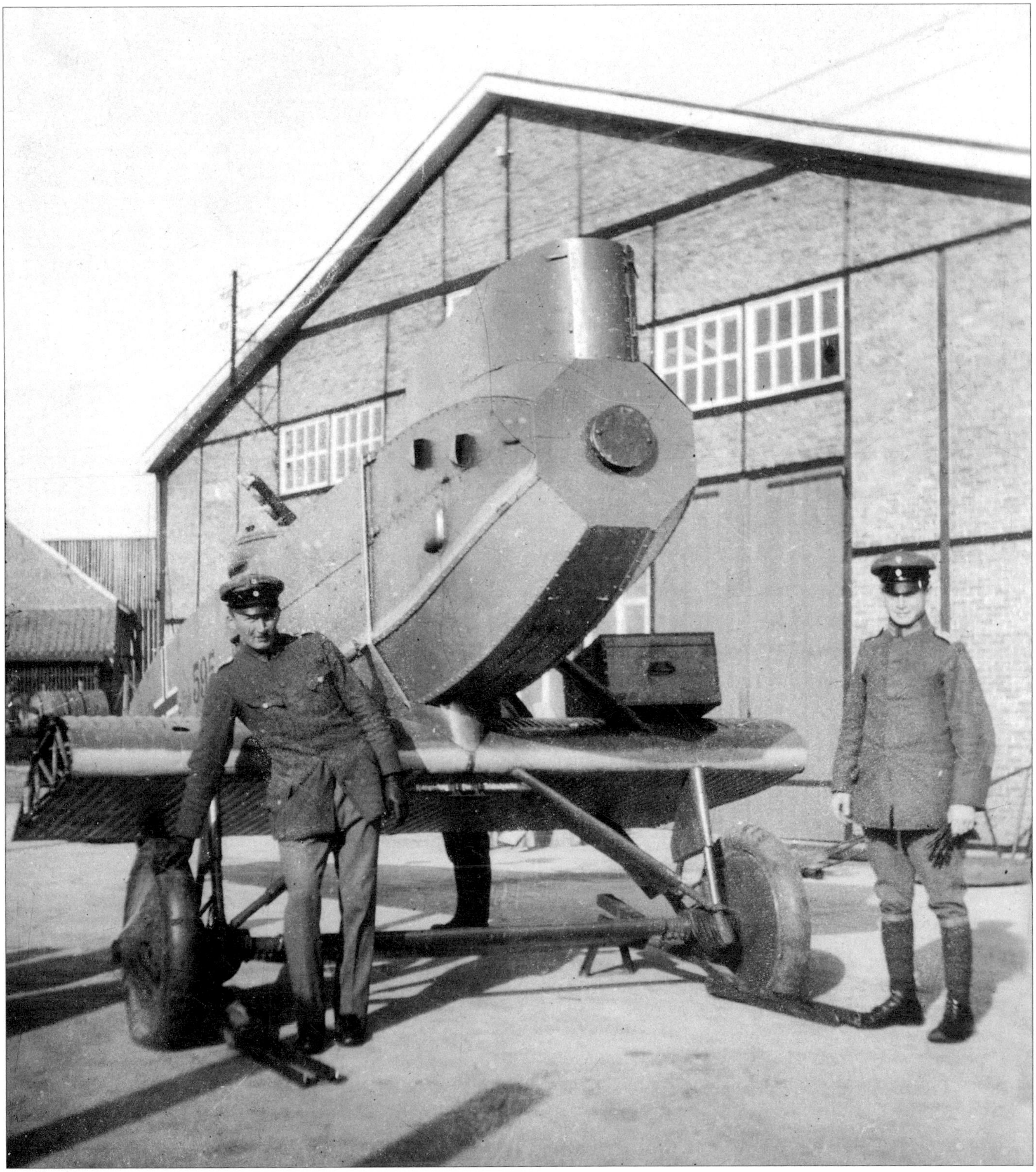

Above: These officers appear to be with brand-new J.I 595/18. The machine has late narrow crosses and the lower surface colour demarcation is very straight. Again, the instructions to treat the corrugated skin with respect are ignored by the large wooden spare parts box placed on the lower centre-section. Shipped, 25 September 1918. (Piotr Mrozowski)

longerons arranged to terminate in a vertical knife edge. There was no internal bracing wires in the rear section such as were in the Fokker D.VII that also has a tubular fuselage. Vertical and horizontal cross tubes with diagonal stress tubes completed the light but strong structure. A steel sleeve was

Above: J.110/17 appears to have suffered little damage in this landing accident. Interim type national markings with thick crosses on the lower surface of both upper and lower wings. Note the difference in tone between the white cross outline and the underside colour. (SDTB)

fitted to the longeron at the juncture with the cross tubes, the latter were flattened at their extremity and riveted to shelves welded to the clip. The sleeves appear to have been universal in nature as they all have attachment points whether used or not. The vertical tubes in the last two bays before the stern-post were replaced by strong bulkheads of duralumin sheet. The longerons were not of the same diameter throughout their length. At the fourth bay (measuring from the armoured nacelle) the longerons were spliced in a similar manner to the wing spars. The two sections of the fuselage were joined by means of substantial steel bolts that are attached to the longerons. The longerons butted up to the rear armour plate at the observer's cockpit. A substantial steel bolt passed through the nacelle and was bolted with a castellated steel nut in the observer's cockpit with the usual split pin.

The exterior of the machine is nearly flush with only the rivet heads and some vents protruding. There are no compound curves but many with single curvature. The forward fuselage weighs 470 kg.

The designers had eliminated control cables wherever possible throughout the machine. The J1 uses a standard control stick, not a wheel, that has a unique feature of having a pilot adjustable double acting spring counterbalance. This was deemed necessary due to the weight of the metal ailerons. Standard spring pre-load is 2.5 kg. The ailerons were operated by means of a steel torque tube "as in the Nieuport scout." The torque tube spans the upper centre section and connects to the ailerons close to the wing-to-centre section attachment points. The torque tube is operated via push-pull linkages. The elevators were operated by stranded cables attached to vertical levers which are actuated by a steel tube direct from the control column. The rudder control employed the customary cables and pulleys. Two pairs of king posts were attached to the rudder and the cables duplicated.

The wings, upper and lower, were built in three sections. The upper plane had a large centre section

Specifications J 4 (Military J.I)					
Source	**1**	**2**	**3**	**4**	**5**
Span upper	16 m	16.76 m	15.0 m	16 m	55 ft 0 in (16.764 m)
Span Lower	11 m				35 ft 7 in (6.223 m)
Wing Chord					
Upper			2.50/2.25 m		8 ft 2 in (2.489 m)
Lower			1.50/1.08 m		4 ft 10 in (1.473 m)
Length	9.20 m	9.04 m	9.06 m	9 m	29 ft 8 in (9.043 m)
Height	3.45 m	3.58 m		3.4 m	11 ft 9 in (3.581 m)
Length aileron					18 ft 0 in (5.486 m)
Chord aileron					1 ft 8 in (0.508 m)
Span tailplane	5.08 m				17 ft 0 in (5.182 m)
Chord tailplane					3 ft 1 in (0.940 m)
Chord elevator					2 ft 2 in 0.660 m)
Gap			2.00 m		6 ft 8 in (est) (2.032 m)
Track					8 ft 1 in (2.463 m)
Weights					
Empty		1,766 Kg	1,766 Kg	1,766 kg	
Load		410 Kg	410 Kg	410 kg	
Total		2,176 Kg	2,1716 Kg		
Performance					
Max Speed			155 km/h		
Climb to					
1,000 m			12.5 min		
2,000 m			33.8 min	32 min	
3,000 m			77.0 min		

Source:
1. Junkers three view 1774. Archives of Technology AP00071.
2 H. Woodman data.
3 P.M. Grosz data.
4. Neuman, G.P. *Die Deutschen Luftstreitkrafte im Weltkriege*. Also in tabulation published in *Illustrierte Flug-Welt*, Vol.1, 1919, P.376.
5. *Report on the Junkers Armoured Two-Seater Biplane Type J.I*, Ministry of Munitions, UK.

with the rear cut-away over the pilot's cockpit and was mounted without dihedral. A system of steel tube struts with aluminium fairings braced the centre section to the fuselage and the lower wing centre section. The leading edge was swept back while the trailing edge was at right angles to the centreline. The lower planes followed the same plan but had about one third the wing area and had dihedral. The lower centre section was built up as one unit with the robust undercarriage. The fuselage was fixed to this unit by short steel tube struts and an aluminium fairing was built around the part of the fuselage that was adjacent to the centre section's upper surface. Two lugs fitted to the bottom of the octagonal armoured forward fuselage were bolted to corresponding lugs welded onto steel sleeves which were in turn riveted to two upper duralumin spars of the lower centre section. The wings were covered with corrugated sheet, lap riveted to the spars at about 35 mm intervals. The wing structure was such that a bullet would not cause damage that would endanger the safety of the whole. A compass was carried in the port lower wing panel.

The British Ministry of Munitions report makes

Above: Assembling J.I biplanes at the Junker's factory. Note the upper wing centre-section on the jig with the J.I fuselage behind. The rear fuselage is covered in four- and five-colour printed lozenge fabric. . The fuselages are marked 873 and 897. The machines still have to have their camouflage and serials marked thereon. These machines were sent to the front from July to October 1918.(SDTB)

Above: Line up of J.I biplanes. These machines carry the interim thick crosses. This is the original version of the heavily-retouched photo on page 107.

As a person Professor Junkers had been well known for his then radical left-wing views and as such he was not wholly trusted by the Military establishment. He was a true pioneer aero dynamist with a tendency to continue to research to solve every problem rather than rush into production. Junkers had given no thought to mass production and the ease of manufacture with the result that the J 4 took four times as many workers as conventional wooden construction.*

The problems with the production of the J 4 were not entirely Junkers fault but he did not have experience with serial production. The military forced him to accept Anthony Fokker into partnership. In theory it looked as if each would balance the other; Junkers with his brilliant mind and Fokker with his experience in mass production of aircraft.

On 20 October 1917, *Junkers Fokker Werke A.G. (Jfa)* of Rötheneer Strasse, Dessau, came into being with a capital of 6,000,000 *Reichmarks* equally contributed by Fokker and Junkers. The patent rights of Junkers inventions were assigned to *Jfa*. However, Junkers' original company *Jco*, as well as the *Forschunganstalt* (Research Institute) remained under his direct control. Both *Jfa* and *Jco* built Junkers monoplanes and Junkers fought for orders for his *Jco* facility with it producing 27 D.I fighters against 13 from *Jfa*!** (See Sidebar for view of *Jco* on the production of the D.I).

Fokker aircraft continued to have the "Fok" stencilled with the serial on all of his aircraft while the Junkers had "Jfa-Fok" stencilled on them, indicating that the Fokker and Junkers-Fokker were two separate entities.

From small beginnings in 1914 the Junkers works grew in size as follows:

Date	Floor Area
January 1915	200 m²
August 1915	650 m²
January 1916	1,850 m²
July 1916	3,800 m²
April 1918	10,700 m²
November 1918	20,400 m²

An airfield for testing was built next to the factory.

The staff increased as the actual construction of aircraft commenced in June 1915 as follows:

Date		Workers	Foremen
June	1915	10	4
January	1916	90	30
April	1917	300	65
April	1918	790	146
November	1918	1,060	190

The designers were Professor Junkers and *Dr Ing* Mader. Series production was executed by the Technical Director of the Junkers-Fokker Works AG, *Dipl Ing* Karl Bauer.***

Notes:

* Morrow Jr, J.H. German Air Power in World War I, University of Nebraska Press, USA, 1982. P105.

** Figures are from Grosz and Terry.

*** Figures are from the ZAK Report.

Comment: *Dr Ing* was the German equivalent of an engineering doctorate degree. *Dipl Ing* was the German equivalent of a diploma in engineering.

continual reference to the "Fokker principle" with respect to the thick wing section with sketches showing how the Fokker D.VII section differs from that used by the J.I. It is therefore obvious that they did not realise that Fokker obtained his introduction to the thick wing from his association with Junkers.

The wings consist of multiple alloy tubular spars interconnected with a web of cross bracing tubes or channels. The upper plane contained ten main spars of 40 mm duralumin tube without counting the aileron tube. These spar tubes were braced from each other by a lattice work of small duralumin tubes, held together by welded steel brackets. The wing covering was 0.3 mm (0.015 inch) thick corrugated duralumin sheet laid parallel to the line of flight. The corrugations are 50 mm centre to centre with a peak to valley depth of 8 mm. The strips were 86 cm wide riveted together with lap joints. The leading edge of the wings is a heavier, 0.8 mm, smooth curved sheet with internal alloy stiffening webs. The alloy sections are joined to the welded steel brackets by rivets. The outer wing sections tapered towards the tips. The upper wing tapered from about 420 mm at the centre section to 150 mm at the tip. The lower wing tapered from about 300 mm to 100 mm. The diameter of the tubes decreased towards the tip and they were carried through by telescoping the pipes into each other with steel collars.[18]

The wings are attached via machined steel threaded couplings attached to the ends of each tubular spar. The same fitting style is utilised for upper wings, lower wings and the stabiliser. The couplings are of two types; male couplings with an external thread and female couplings with an

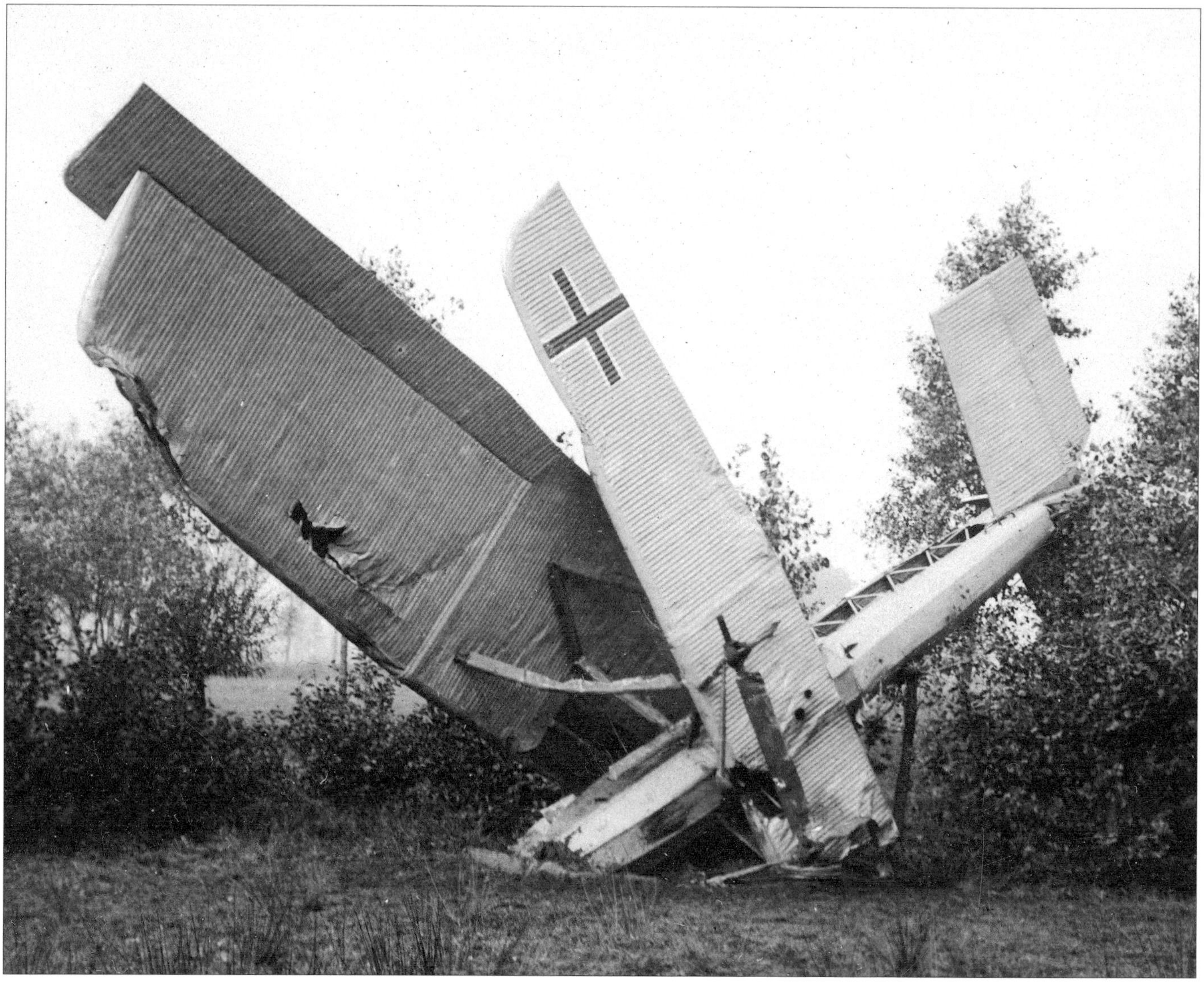

Above: Under surfaces are well shown by this crashed J.I biplane. The difference in the tonal values of the underside of the upper wing from that of the lower wing cannot be due to lighting alone; it is possibly an effect from previous maintenance. The white edging for the national marking is different from the under-surface colour. (Piotr Mrozowski)

internally threaded locking ring. Male fittings are inboard and female fittings outboard i.e. outer wing panels. The steel fittings are attached to the alloy tubes with steel rivets. The male coupling has a tapered flange outboard of the threaded portion. This fits into a reverse tapered socket on the female coupling. This taper fit provides exact positioning of the parts and the internally threaded locking ring engages the external thread on the male coupling to secure the two sections together. On J.586/17 the locking ring has drilled holes around the periphery that, while designed for a specialised wrench, show evidence of having been secured with a drift and hammer. There is limited access to the coupling when assembled and the drift and hammer method could have been a field applied expedient.

The trailing edge of each panel terminates in an evenly radius scallop. This looks like a stamping as opposed to a rolled corrugation as the trailing edge is flat and the height of the valley in the corrugations. The leading edge of each upper corrugation is gathered together where the wing panels terminate under the leading edge. The wing covering panels are uniform in width and are one piece front to back. The outboard edge of each panel overlaps the next by approximately 20 – 25mm. Panels are fastened to each other with evenly spaced alloy rivets.

Of note is that, while the exterior of the wing was painted with a camouflage pattern the interior of the wings appeared to have a flawlessly applied clear protective coating. This appeared to be either a skin manufacturer's machine applied coating or very

Above: Australian airmen from No.4 Squadron at Bickendorf airfield pose with a German airman and one of the wrecked J.I biplanes at this airfield, J.884/17. Left to Right: Ltn Gentz (German airman); Major Ellis, MC; Capt Roy King, DFC; and Capt Jones. The machine appears to be in good condition apart from the souvenired fuselage cross. (AWM E04146)

good spraying on the production floor. There was no evidence of any brush strokes. Compared to the dull finish on the bare alloy structural tubes, the inner skin was smooth and shiny.

Balanced ailerons were fitted to the upper wing only and carried the thick wing profile to completion. Each commenced at the centre section and ran the full length of the outer wing panel. A hinged aerodynamic gap filler strip about 100 mm wide and capable of upward movement only was fixed at the rear edge of the upper plane thus bridging the gap between the aileron and the wing. This strip was made of the same corrugated material as the wings. The strip relied on airflow to stay in place and has a piano wire hinge. No cables were used in actuating the ailerons. A Junkers radiator was carried on the leading edge of the upper centre section (5 ft 8 in x 7 ft 3½ in). The compass was carried on the left hand bottom plane.

There were three groups of struts the arrangement of which was considered "one of the most interesting features of the Junker biplane." There were no interplane struts. Two struts connected the lower plane to the upper plane. A second pair connected the upper plane to the lower longeron of the fuselage while another pair connected from here to the lower wing panel to which the undercarriage struts were connected. All these struts were of steel tube with aluminium fairings. They were joined to the spars by riveted steel collars carrying welded-on lugs. At the fuselage, the fork ends of the struts were bolted to the lugs welded to the armour plate. The struts were of fixed length with no facility for adjustment. The undercarriage was built into the lower centre section. It comprised two vees of normal design except that an additional tube was welded into the upper extremity of the forward limb to a point just below the middle of the rear strut. The vees were parallel and vertical. The undercarriage was sprung with two coiled springs and a double loop of cable limited the upward travel of the steel tube axle. Wire bracing was unnecessary.

The unique J.I was the subject of many photographs that appear in private albums of members of the various aerial services who undertook occupation duty in Germany after the Armistice. Those at Bickendorf aerodrome, near Cologne, reappear time and again along with the wrecked *Risenfluge* wrecks on that airfield. The following photographs show J.884/17 as it deteriorates through the winter of 1918–1919. It is not known how many J.I biplanes were at Bickendorf but the photographs appear to show more than the one machine. Some of these photographs are not of the best quality but they present an unique opportunity to show the situation on this airfield under Allied control.

Above: This J.I displaying the internal construction of the engine access door appears to be another J.I. The attending servicemen are the same. (AWM)

The tail skid was of ash hinged to an inverted pylon and internally sprung. The skid and the engine bearers were the only wooden components in the structure. All fittings were attached to wood mounts that were in turn bolted to the armoured shell.

The rectangular tailplane was of similar construction to the wings with seven tubular spars and corrugated metal covering. Divided elevators were fitted. A large fixed metal fin was braced to the fuselage with a balanced metal rudder completing the tailplane. From J.850/17 onward the rudder and fin were covered with corrugated metal.

The task of reconnaissance was considered of more importance than arming the J.I with a synchronised machine gun for the pilot, indeed, at the altitude that the machine would be operating at, it was unlikely it could be used effectively. A Telefunken W/T sending/receiving set was considered more important. The equipment was powered by a strut mounted wind driven generator. The observer/gunner had the usual Parabellum LMG 14 machine gun on a ring mount. Three ammunition drums were fitted to the right and left of the observer's seat; with a further two drums under the seat. From J.101/17 machines were fitted for carrying cinematograph hand held cameras and magazines. Racks for 10 signalling flares were mounted on each side of the fuselage.

Some sources state that some aircraft had provision for two Parabellum guns to fire downward

Above, Below, & Following Two Pages: J.884/17 early in its association with the Allies at Bickendorf. The fabric is intact and the machine is in overall good condition. Note the biplane tail of a Friedrichshafen G.IIIa bomber in the foreground. The starboard fabric appears to have been later removed to souvenir the whole serial and national marking.

Junk J.I.
884

The Page: Hulk of an engineless J.I at Bickendorf. That the Junkers was the subject of intense interest is shown by the number of photographs taken by different Allied airmen who were at Bickendorf on occupation duty post-Armistice. The wreckage that accumulated at the aerodrome is visible in the photographs. The tail of a Friedrichshafen bomber and various wings may be seen.

Above & Below: Young British airmen poses with abandoned J.I wrecks probably at Bickendorf. Note the wing damage and location of the second machine that may well be J.884/17. (AHT AL0447-030 & 031)

Above: Probably Bickendorf. "All metal Junkers-Mercedes 180".

through holes in the fuselage nacelle. When these guns were carried the W/T had to be removed. The Allies reports noted that there was "no trace of guns having been fixed in the pilot's seat, or on the floor of the observer's cockpit firing downwards, as on the A.E.G. two-seater."[19] This is in keeping with the main task of the J.I, reconnaissance, not trench strafing. Terry Gerard wrote that contemporary Junkers records in his possession noted that a *Ltn* Pfeffer tested a J 4 with two fixed, downward firing machine guns for trench strafing. They were mounted to the left of the observer/gunner. In early 1918 *Idflieg* requested provision for guns rather than W/T equipment.[20] While it seems that this was not carried out on the production J 4 machines and these installations remained experimental, observations by Allied personnel at Bickendorf airfield state that the Junkers did have downward firing guns.

Serial Allocation

Serial	No. Ordered	Notes
J.425/17 to 427/17	3	18 Nov 1916. Prototypes.
J.100/17 to 149/17	50	March 1917. First production batch.
J.800/17 to 899/17	100	December 1917.
J.576/18 to 615/18	40	20 June 1918.
J.716/18 to 725/18	10	12 July 1918
J.726/18 to 805/18	80	Oct 1918. (Anderson & Sanger record 726/18 to 825/18.*)

* Anderson & Sanger record aircraft ordered and delivered after the Armistice.

Specifications J 4 (Military J.I)

Source	1	2	3	4	5
Span upper	16 m	16.76 m	15.0 m	16 m	55 ft 0 in (16.764 m)
Span Lower	11 m				35 ft 7 in (6.223 m)
Wing Chord					
Upper			2.50/2.25 m		8 ft 2 in (2.489 m)
Lower			1.50/1.08 m		4 ft 10 in (1.473 m)
Length	9.20 m	9.04 m	9.06 m	9 m	29 ft 8 in (9.043 m)
Height	3.45 m	3.58 m		3.4 m	11 ft 9 in (3.581 m)
Length aileron					18 ft 0 in (5.486 m)
Chord aileron					1 ft 8 in (0.508 m)
Span tailplane	5.08 m				17 ft 0 in (5.182 m)
Chord tailplane					3 ft 1 in (0.940 m)
Chord elevator					2 ft 2 in 0.660 m)
Gap			2.00 m		6 ft 8 in (est) (2.032 m)
Track					8 ft 1 in (2.463 m)
Weights					
Empty		1,766 Kg	1,766 Kg	1,766 kg	
Load		410 Kg	410 Kg	410 kg	
Total		2,176 Kg	2,1716 Kg		
Performance					
Max Speed			155 km/h		
Climb to					
1,000 m			12.5 min		
2,000 m			33.8 min	32 min	
3,000 m			77.0 min		

Source:
1. Junkers three view 1774. Archives of Technology AP00071.
2 H. Woodman data.
3 P.M. Grosz data.
4. Neuman, G.P. *Die Deutschen Luftstreitkrafte im Weltkriege*. Also in tabulation published in *Illustrierte Flug-Welt*, Vol.1, 1919, P.376.
5. *Report on the Junkers Armoured Two-Seater Biplane Type J.I*, Ministry of Munitions, UK.

Junkers J.I at the Front

	1917				1918			
At end of	Jun	Aug	Oct	Dec	Feb	Apr	Jun	Aug
No.	0/7	1/2	1/12	4/76	16/113	25/139	25/184	60/186

The above chart shows the numbers of J.I biplanes at the Front against the total number of J Types (A.E.G. and Albatros J.I and J.II types). The longer construction times required by the Junkers aircraft meant that they would always be in the minority. In November 1917 the Dillinger foundry was flooded, preventing the supply of armoured plate for Junkers machines. Source: Morrow Jr, J.H. *German Air Power in World War I*, University of Nebraska Press, USA, 1982. P104.

Above & Below: J.884/17 continues to deteriorate as the snow melts away. Note the start of damage to the lower port wing and missing exhaust pipe.

Above: The fabric with the serial appears to have been cut away but the location and damage as well as the camouflage on the corrugated fin suggest that this is J.884/17.

Sqn Ldr George Jones, while at the RAF Staff College, Andover, was one such person. He served with No.4 Squadron, AFC, and was at Bickendorf aerodrome, Cologne, after the Armistice. Discussing ground strafing Jones noted that

it was interesting to find later, at Cologne, a "Junkers" Aeroplane which had been designed and used for the attack of ground targets. It was all

Above: J.884/17 in the same location with bomber wings deposited to the left of the photograph.

metal; the wings and fuselage being covered with finely corrugated aluminium and from the nose to behind the rear cockpit with sheets of steel. It was armed with one gun firing forward, two through the bottom of the fuselage, and one concludes, a movable gun for the rear cockpit. It gave the impression of being too heavy to fly

The trenches were not laid out in a straight line and to attack them the aircraft would have to fly low and be continually changing direction to bring the guns over their target, not to mention the difficulty of aiming the guns. The main purpose for the J-Types with this type of armament was to attack troop concentrations in an advance, not the trenches themselves. The attacking machines had to fly low within the range of small arms ammunition and shrapnel. Whatever the armament, it was very dangerous work and many observers wore the standard infantry steel helmet rather than the leather flying helmet.

The 2 cm Becker Type 3 canon was carried by some aircraft by September 1918. The Becker cannon was comparatively light but required a strong mounting and ring. Some J 4 biplanes were equipped with this gun before the end of the war.

The J 4 was issued in one and two to the ordinary *Fliegerabteilungen* units commencing in from August 1917. It was given the nick-name *Möbelwagen* – Furniture Van – by pilots who were used to the lighter two-seaters.

What was thought of the J 4.

According to the J.I handbook the loss of a wheel caused a machine to turn over but only the tubular framework of the wings was damaged, the machine not being wrecked completely.

The handbook also stated that, despite being made of metal, the strength of the metal covered wings while quite sufficient for normal flying loads, could not carry a heavy load over a small area. Care had to be taken not to stand or lean on the wings nor place any heavy article with sharp edges, such as petrol tins, upon them. "Never move the machine by pressing on the struts."

The machine could be erected with six to eight men in four to six hours. The fact that no rigging was necessary was one factor in the short time required to erect the J.I.

Major repairs could only be carried out by specially trained personnel. A tool kit was provided that allowed small repairs to be carried out in the field. Damaged struts were to be entirely replaced and riveted to the connecting pieces in the usual way. Damaged undercarriages were to be replaced with new parts.

The heavy J.I required a long and firm airfield. The *Flieger-Abteilunge* crews found the machine had relative good manoeuvrability, agreeable landing qualities, good protection, such that the excessively long take-off run was not considered a detriment. In a war of movement, when fields were changed almost daily, the Junkers was often left behind.

The British were aware of the Junkers armoured biplane as a report in the *Daily Chronicle* at the end of November 1917 noted that the new "Junker" machine with an all-metal body was in production for accompanying the infantry in advance.[21] An official report for January 1918 on Enemy Aviation

Above: "Raymond and Uphill on Junker." Members of No.48 Squadron on a J.I biplane. (AHT AL0498-011)

Above & Below: British officers and men of No. 48 Squadron pose with derelict J.I biplanes. Note the S.E.5a of No.29 Squadron (single white band around fuselage near tailplane) in the background would seem to indicate that this was Bickendorf aerodrome. (AHT AL0498-016 & 026)

Above: Another abandoned J.I near a railway line.

Above & Below: An officer and his ground crew from *Fl Abt* 7, posing with their J.I on a field that boasts a substantial wooden hangar. The machine is in the 800 range and has the narrow late war cross national insignia. (AHT AL0175-025)

Junkers & Co. Dessau

Above: J.146/17 in two-colour camouflage scheme. A rack containing flares is situated on the fuselage at the observer's position. The aileron actuating controls leading into the cockpit are clearly shown. Note the "*Jfa*" logo on the wheel covers. The thick interim cross outlines extend around the leading edge of the lower wing. The Axial logo may be seen on the airscrew. The fin is fabric covered but he rudder is corrugated metal. On the original photograph the stencil on the upper wing leading edge can be read: Junk. I J. 146. Note also the dark colour that the radiator is always painted.

Below: J.118/18 of *Fl Abt* A208 has a camouflaged rudder. Note how the corrugated sheet covering the aileron gap stands proud of the wing. The airflow would force the sheet down when in flight.

Facing Page: Assembling a J.I in the field. Unfortunately, the serial is too faint to read in the stencilling on the nose. Note the wheel cover stencilling.

Above: Recovering a J.I in what was unfortunately a not uncommon landing position. This machine has the camouflaged rudder and early insignia. The cross under the upper wing does not appear to have any white surround.

Above: The crew of J.818/17 with well-wishers after coming into conflict with the field boundary, Note the damage to the upper starboard wing. The lower starboard wing tip has been sheared off. Again, the rugged J.I structure has held together. The machine is camouflaged in two-colour scheme with the fuselage cross marked on a light panel. Note the streamer hanaging from the starboard lower wing tip.

Above: This late J.I with corrugated fin is in full late war colours with the armoured forward fuselage camouflaged in two colours and the rear fuselage covered with lozenge fabric. The wheel covers have the Jfa logo but no serial is stencilled onto the fuselage. The substantial building in the background may mean that the personnel sitting with the machine are at the Junkers factory. The serial number was overpainted by the censor.

Above: This abandoned J.I was possibly photographed at Chimay, Belgium. (AHT AL0076-090)

Above: A British airman examines the armour of a wrecked J.I post Armistice. (AHT AL0613-014)

Above: An American serviceman stands guard over four J 1 biplanes. 576/ and 596/ appear to be the first two in the line-up. (SDASM)

Above: American personnel examine Junkers J.I fuselages. Note that they appear to be all painted the same dark colour.

Above: American servicemen posing in front of a Junkers J.I. The fin corrugations were parallel to the bottom of the fin and the cap strip gave the appearance of a slope to the upper edge of the fin where the corrugations were cut.

Above: This J.I fuselage has "Unserviceable" chalked on its nose. Note the factory mounted flare box near the observer's cockpit and the chalked number '144' and partially visible pilot harness. .

Above: J.I fuselages in US hands at Romorantin, post-Armistice. The fuselage chalked #104 is J.846/17. According to US documents J.846/17 had two forward-firing slanted Spandau machine guns in the observer's cockpit. (SDTB)

Above: Another J.I fuselage at Romarantin. The hangar bays appear to be full of AEG G.IV bomber fuselages (one with with a white 'Icarus' personal insignia painted on the nose) and parts.

Above: More J.I fuselages obtained by the U.S. after the Armistice.

Above: US personnel with a reasonably intact J.I. The machine appears to be in plain metal finish but on close examination camouflage paints are peeled off.

Junkers J.I J 115/17

Underside of upper wing, J.I 115/17.

Above: A J.I with the wings detached and stored against the fuselage post-Armistice. (SDASM)

Above: This may be the same aircraft, note the lack of a drive shaft from the engine. This photograph appeared in many British publications, usually with an incorrect caption such as this from a Canadian post-war publication:- "This German Aeroplane, brought down behind the British lines, is an object of interest to these Canadians. Its wings, made of corrugated aluminium and practically fire and weather proof, were a German invention, and enabled it to be left for long periods in the open." Canadian Official photograph dated 21 November 1918.

Above: Allied personnel examining another abandoned J.I. (SDASM)

Above: US personnel examining German aircraft after the Armistice. The J.I fuselage is in front of AEG bomber fuselages that appear to be in good condition. Note the size of the storage box behind the observer's cockpit. (AHT AL0111-145)

Above: US personnel appear to be dismantling this J.I on a snow-covered airfield. The ailerons are not connected allowing them to swing down showing off the balance surfaces. Note the corrugated fin.

Above: J.I 868/17 after suffering engine failure appears to have lost a wheel carrying out a forced landing on 30 July 1918. The machine would have been on factory trials as it does not have camouflage or national insignia applied. The serial is repeated on the nose behind the airscrew. Despite the damage seen here, the machine was repaired and sent to the front in the week of 9 September 1918. (AHT AL0459-215)

Above: A poor photograph but it shows what happens when you hit a horse on landing. Note the horse's body in the foreground of the photograph. (AHT AL0192-31a)

Above: J.813/17 from *Fl Abt (A)* 259 on its nose gives a good view of the upper surface camouflage. Note the proportions of the wing crosses. There appears to have been no standard layout to the two-colour scheme applied to the J.I.

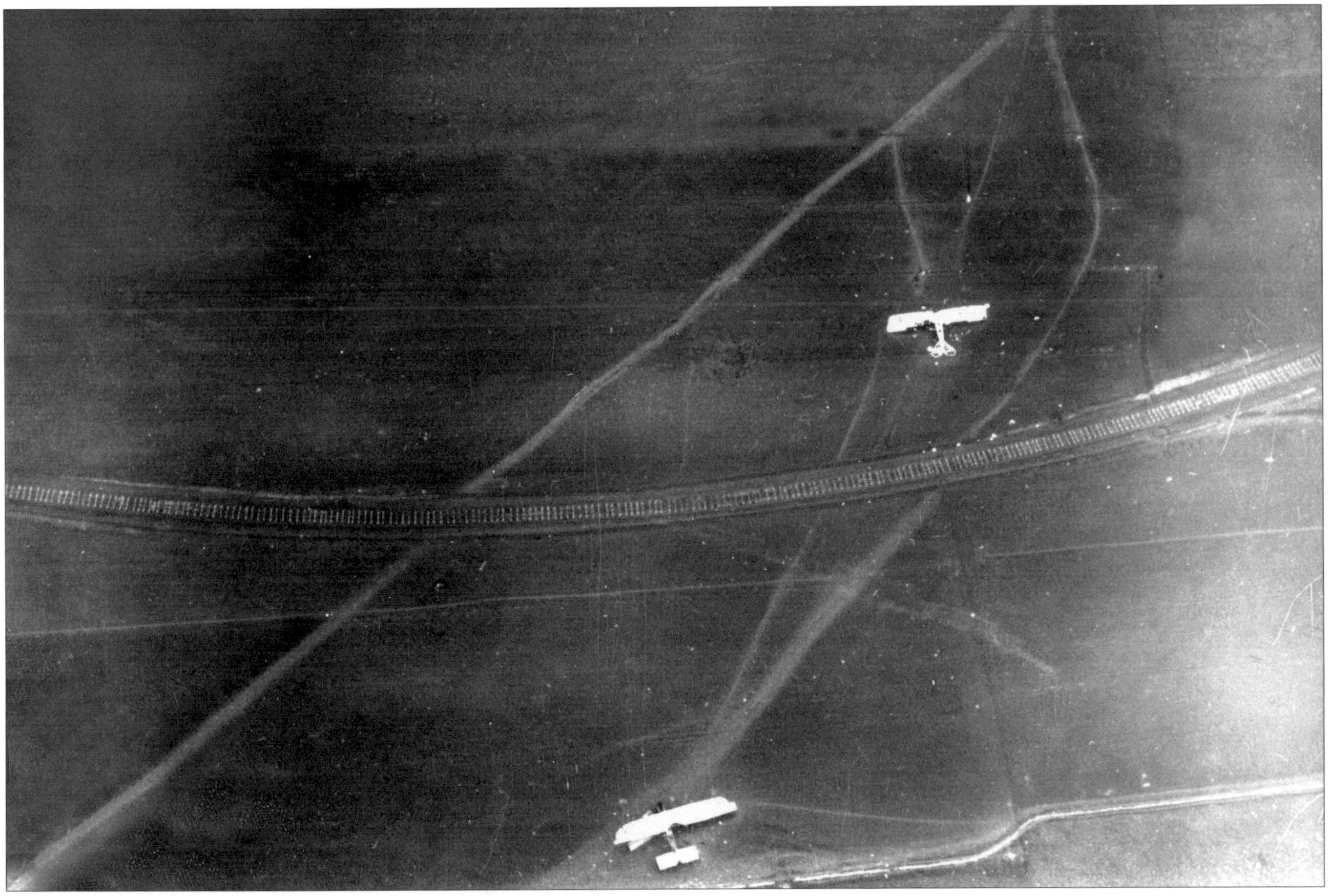

Above: A Junkers J.I in dire circumstances after landing under the railway line that runs across the middle of the photograph. Note the position of the port lower wing. The other two-seater biplane over the railway line is in very poor condition. (SDTB).

Above: J.126/17 must have hit the ground with considerable force. The armoured shell is intact but the landing gear and wings have taken the brunt of the force. Note the style of serial application, the individual number "6" and the position of the national insignia. This machine had been sent to *Armee Flugpark* 17 in the week of 25 February 1918 for acceptance. (SDTB)

Above: The snow covered airfield provides a good backdrop for this pristine J.I biplane's clean lines with the spinner in place. There are national insignias on the underside of both wings. (AHT AL0298-05)

Above: An abandoned late J.I with corrugated fin still retains its camouflage pattern.

Above: J.106/17 displays the early colour scheme where the fin was camouflaged. The style of serial should be noted. The underwing crosses on the upper wing have a white outline.

Above: Note how the single colour camouflage on this J.I appears dark. The airscrew is covered with a canvas cover.

Above: The camouflage on this J.I serial 81X/17 is in broad bands rather than splotches. It has interim crosses with red/white /black streamers on the wing tips. Note the canvas hangar in the background.

Above: There are three exhaust pipes and two fins in this photograph of more J.I fuselages in US hands. Note the armoured fuselage underside is a darker colour than the unarmoured rear fuselage underside. (SDAM)

Above: J.I on a railway wagon undergoing maintenance. The rear fuselage is uncovered and this may be a post-Armistice scene. Note the cable to restrain the open engine door.

Above: This J.I appears to be in very good condition although the rear fuselage fabric is missing.

Above: The wings of this J.I show mistreatment, otherwise it appears to be intact. The dent in the port upper wing identifies this as 586/18, the Canadian war trophy machine. The presence of US personnel indicates that this was taken at Romorantin. (SDASM)

Below: This is the same machine shown above. (SDASM)

Above & Below: Two more views of 586/18 at Romorantin.

Above: The crew of this Junkers pose with their crashed machine. The late national insignia and camouflage pattern is well shown. (AHT AL0613-009)

Above: Despite the caption on this postcard this is another view of Romorantin. (See pages 152 & 153.)

Above: Another view of J.134/17 with complete disregard for treating the metal wings with caution. Note the wheel cover stencils and lack of a spinner to the airscrew.

Above: While *Jasta* 36 was at Saultain airfield in the first ten days of October 1918, these two Junkers J.I biplanes were there as well. J.808/17 is to the rear. The serial of that nearest to the camera could possibly be J.803/17 or J.805/17. Both have a band around their fuselages, wingtip streamers and an instrument attached to the interplane strut. Note how dark the interplane struts are painted.

Above: J.749/18 the so called J.II with corrugated rear fuselage, lightened airframe and new fin and rudder in the Junkers factory late 1918. (SDTB)

Above: J.749/18 on the Dessau field 14 February 1919. If this version of the J 4 had gone into production there is no doubt that it would have received the military designation J.II.

Above: The marking 749 is just visible on the fin. Note the CL.I monoplane in the background. Photograph dated 22 February 1919.

Above: Another view of J.749/18 on the snow covered Junker's Dessau airfield. Winter 1918–1919. (SDTB)

Above: The long wing 749/18. Nothing has been found on the evaluation of the machine with these long-span wings. Note the device under the wheel axles to take the strain off the undercarriage. 22 February 1919.(SDTB)

Above: J.849/17 has the serial stencilled on the armour underneath the cockpit and crudely marked on the corrugated rudder. The fin is fabric covered. The handling trolley has been crudely removed from the photograph but its shape can still be discerned. This aircraft crashed on 24 October 1918.

Above & Right: A production J.I in Allied hands after the Armistice. The serial appears to be J.819/17, but is not clear. The machine is in very good condition. The meaning of the "2224" chalked on the fuselage is unknown. The built-up wing spar and corrugated covering is well shown.

Above: This J.I has suffered badly. The wing covering has been deformed and torn in places. The corrugated metal sheets required a lot of respect as it was not a load bearing surface.

mentions that the "Junkers armoured two-seater also now in use" along with the A.E.G.[22] The following month the A.E.G. and Albatros armoured aircraft "As used for Infantry Contact Work" are reported on but the report does not mention the Junkers biplane.[23] In the March report the A.E.G. is considered the best, however the French report of April 1918 noted that the armoured Junker appeared "to have given excellent results during the recent attacks, but there are few of these machines (R.F.C.) Confirmed by pilots and prisoners. Would appear to be the best for infantry co-operation."[24] An RAF report in May stated that the with regard to the J.I the "blind spot appears to be forward and slightly underneath," indicating that the type had been engaged in aerial combat.[25]

According to Grey and Thetford the J.I was viewed with considerable favour once the revolutionary aircraft's peculiarities were understood,[26] while an American intelligence report stated that the type was unpopular with pilots as being too heavy and unwieldy to be of use.[27]

The following reports show what the men who flew in the J.I thought about their machine. They are presented at length as firsthand accounts such as these are rare.[28]

The following statement was made by *Ltn d.Res* Wagner of *Fl Abt* 268:-

On the 28th March I made a flight over the front at 80 meters height in your machine No.128. During which we received over 100 hits, my helmet was shot thru and my observer was wounded.

I am convinced that this machine saved my life. It is impossible to make this kind of flight over the enemy trenches at this time and at such low height.

The following is a Memorandum from *Armee-Oberkommando* 4 (Army-Headquarters 4), *Kommandeur der Flieger* (Commander of Aviation of an Army), A.H.Qu (Army Headquarters), dated 2 October1917, to the *Kommandierenden General der Luftstreitkräfte* (Commanding General of the air force), GR.H.Qu (General Headquarters)

The attachments presented are reports about the *Junkers-Infanterie-Flugzeug* from *Fliegerabteilung*

Above & Below: Combat damage to the J 4 's structure.

Left: Bullet holes have been marked in the wing structure of a J 1. The machine managed to return safely despite the damage.

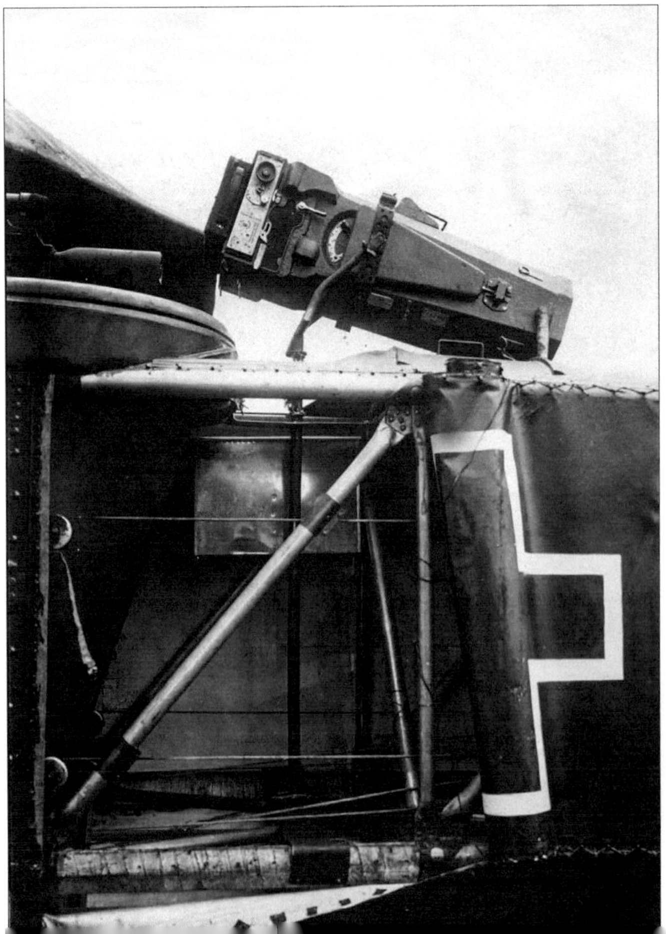

Above & Left: A camera mounted near the Parabellum gun ring on a J.I in 1918. The camera is an ICA *Flieger Kamera* F.K.II f=50 cm Lens: Triplet 1:4,8 (Tessar). Also note the dark (black?) painted fuselage.

Facing Page: J 4 special guns; openings for two machine guns for ground straffing. These holes are in the bottom of the J.I in Italy. The scale reads 8 inches on the top and 200 mm on the bottom. The holes are approximately 170 x 100 mm.

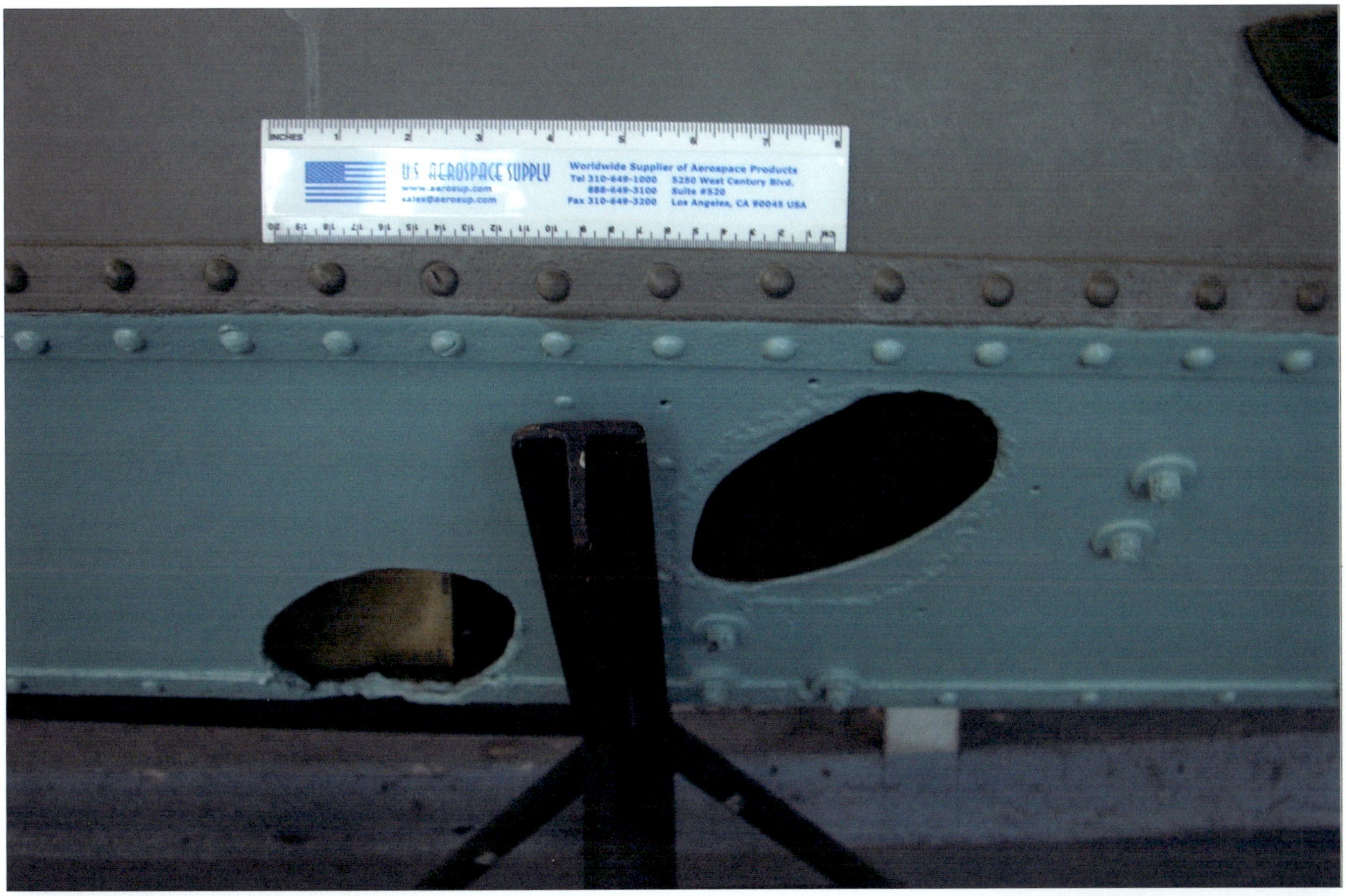
INCHES
U.S. AEROSPACE SUPPLY
www.aerosup.com
sales@aerosup.com
Worldwide Supplier of Aerospace Products
Tel 310-649-1000 5250 West Century Blvd.
888-649-3100 Suite #520
Fax 310-649-3200 Los Angeles, CA 90045 USA

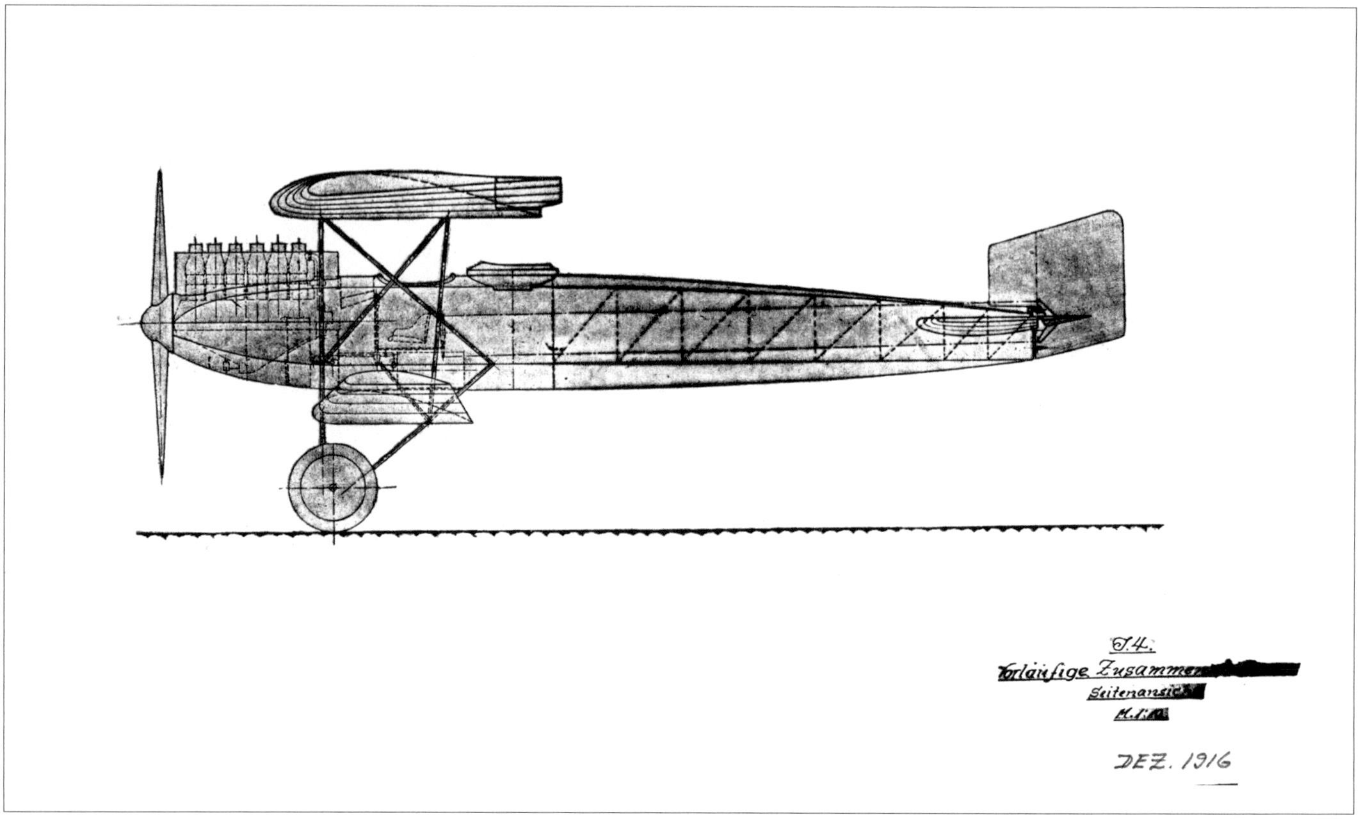

Above: This drawing from December 1916 shows the layout fully developed and a balanced rudder.

Below & Facing Page, Top: These drawings from December 1916, appear to be in the same sequence as the figure above.

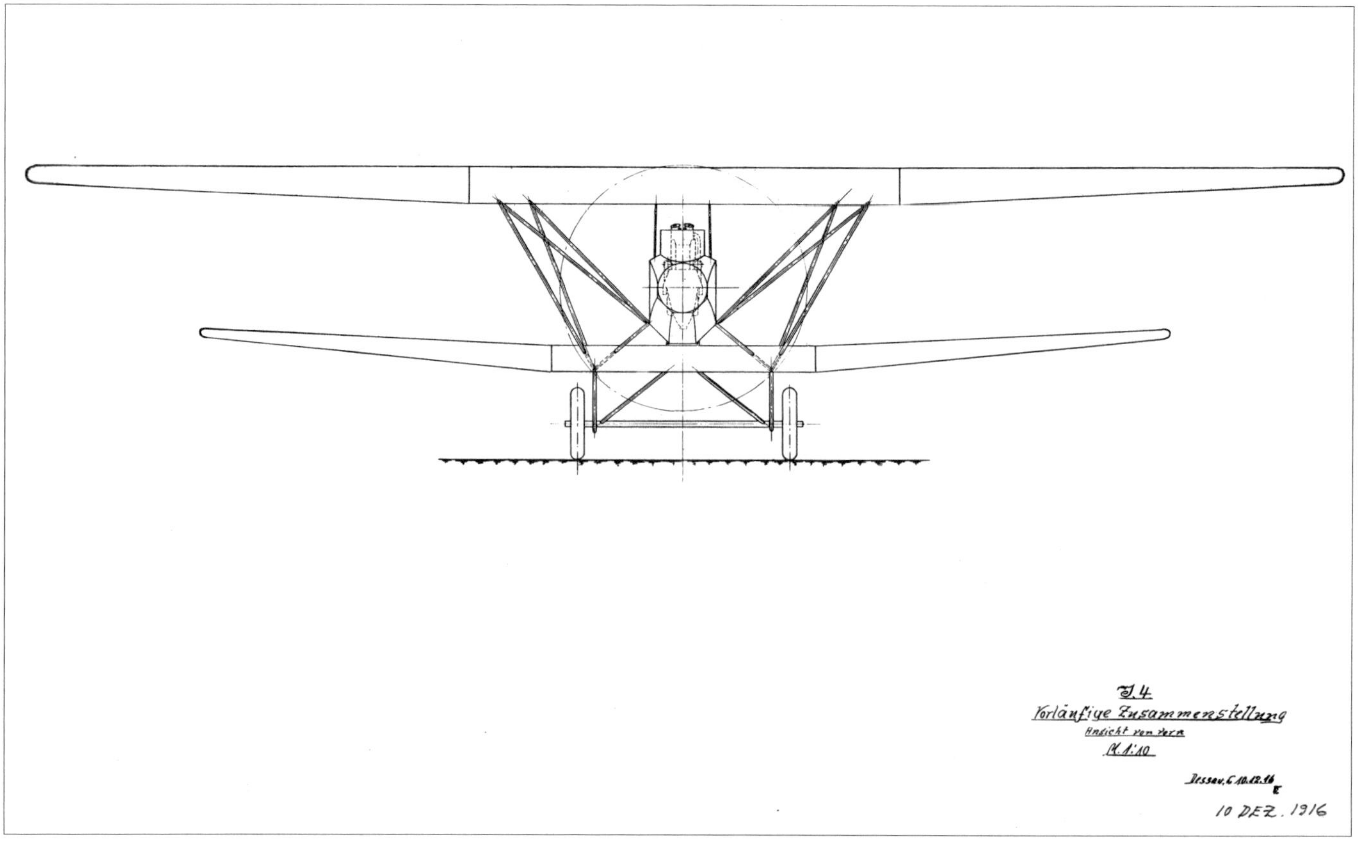

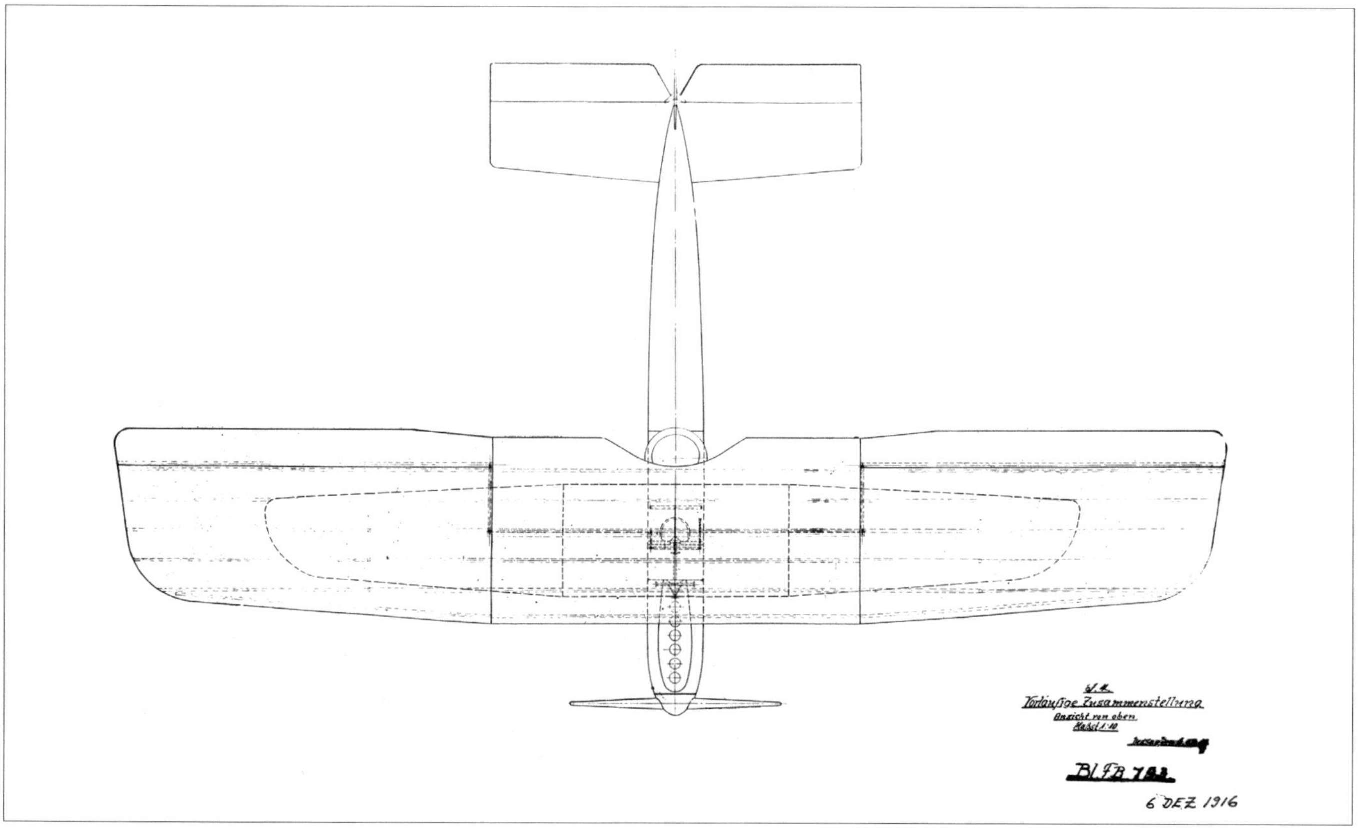

Below & Following Page: Dated 13 March 1917, these drawings show the prototype as it emerged with plain rudder and fin.

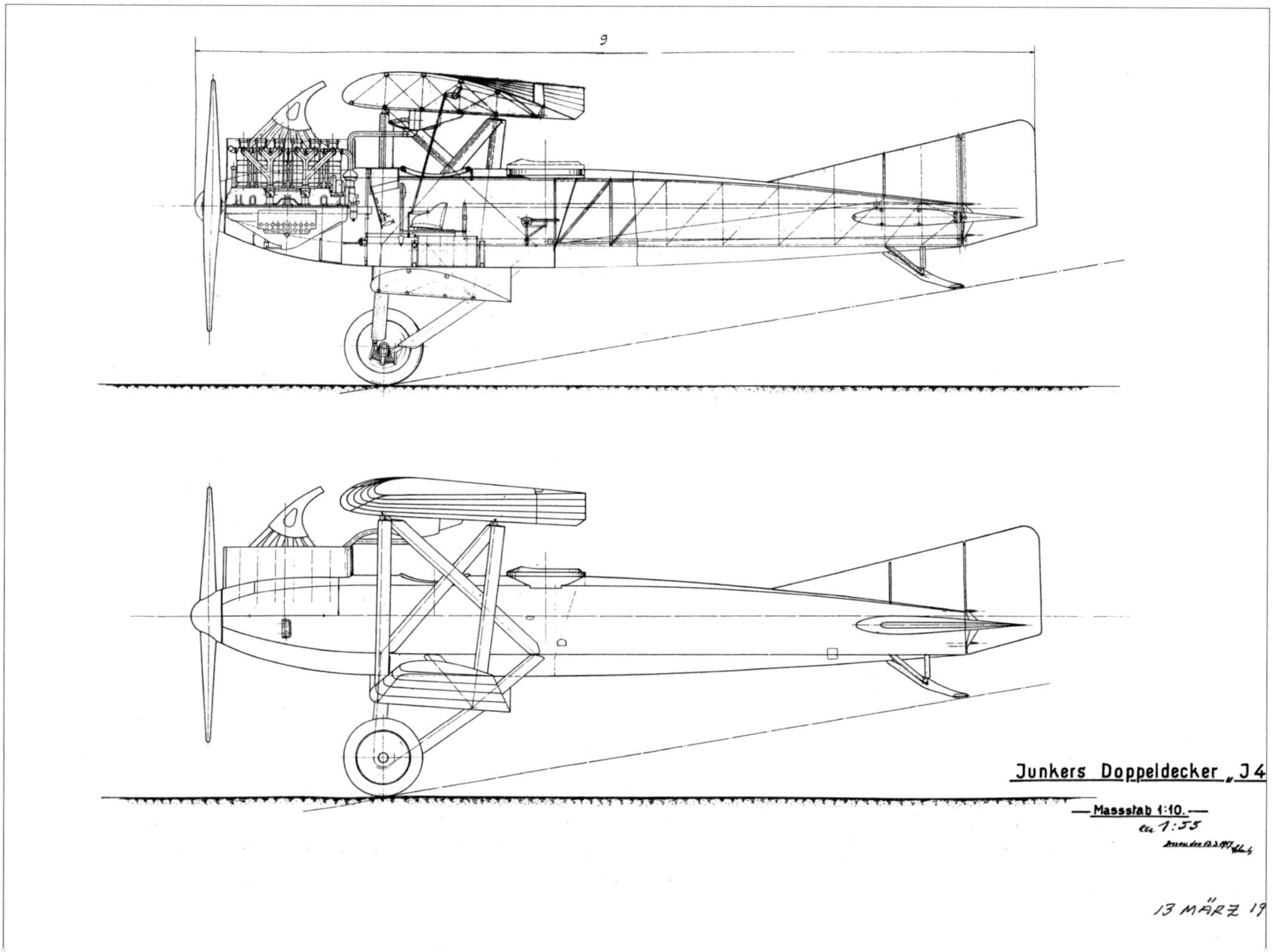

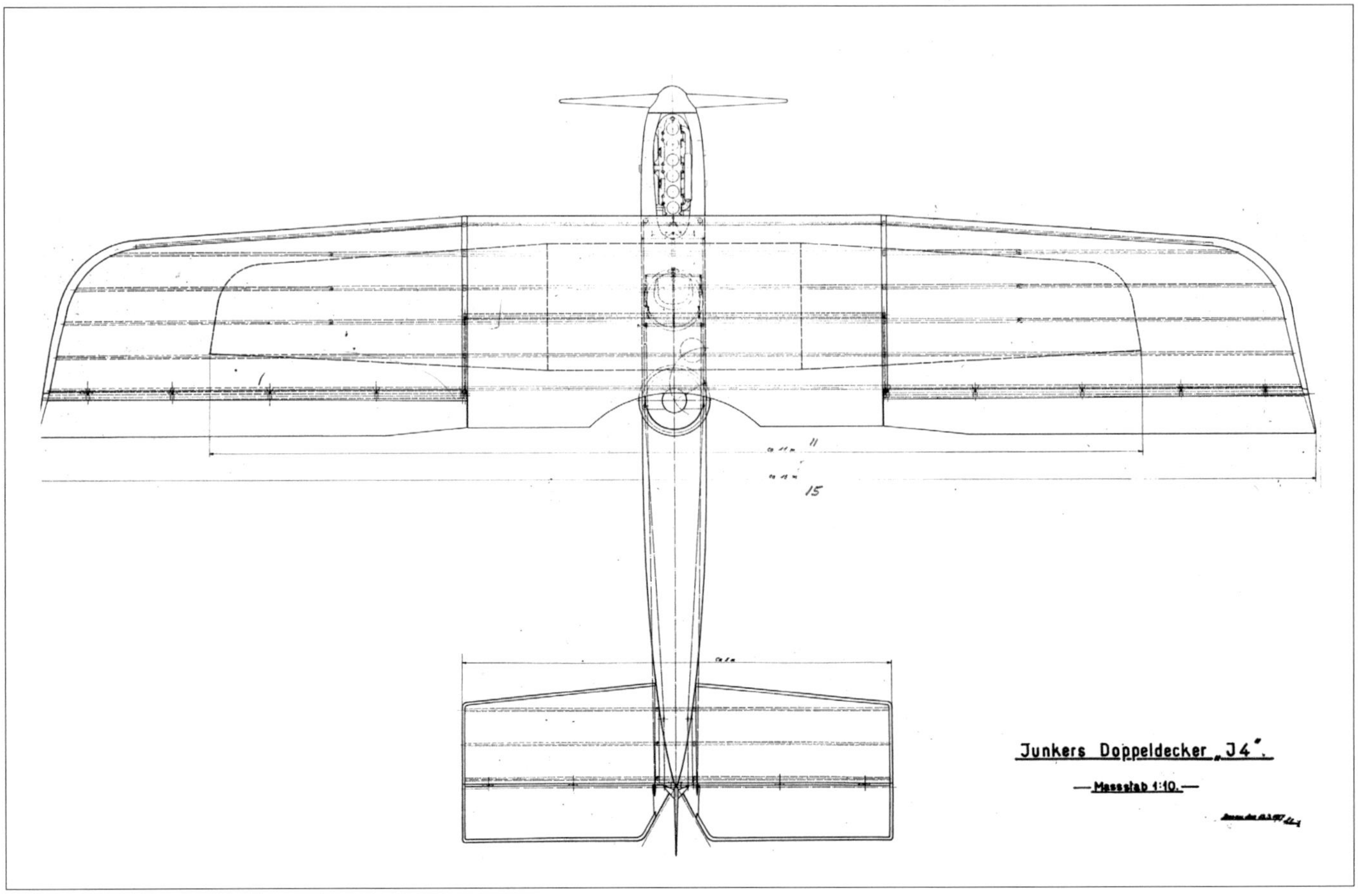
Junkers Doppeldecker „J4".
— Massstab 1:10. —

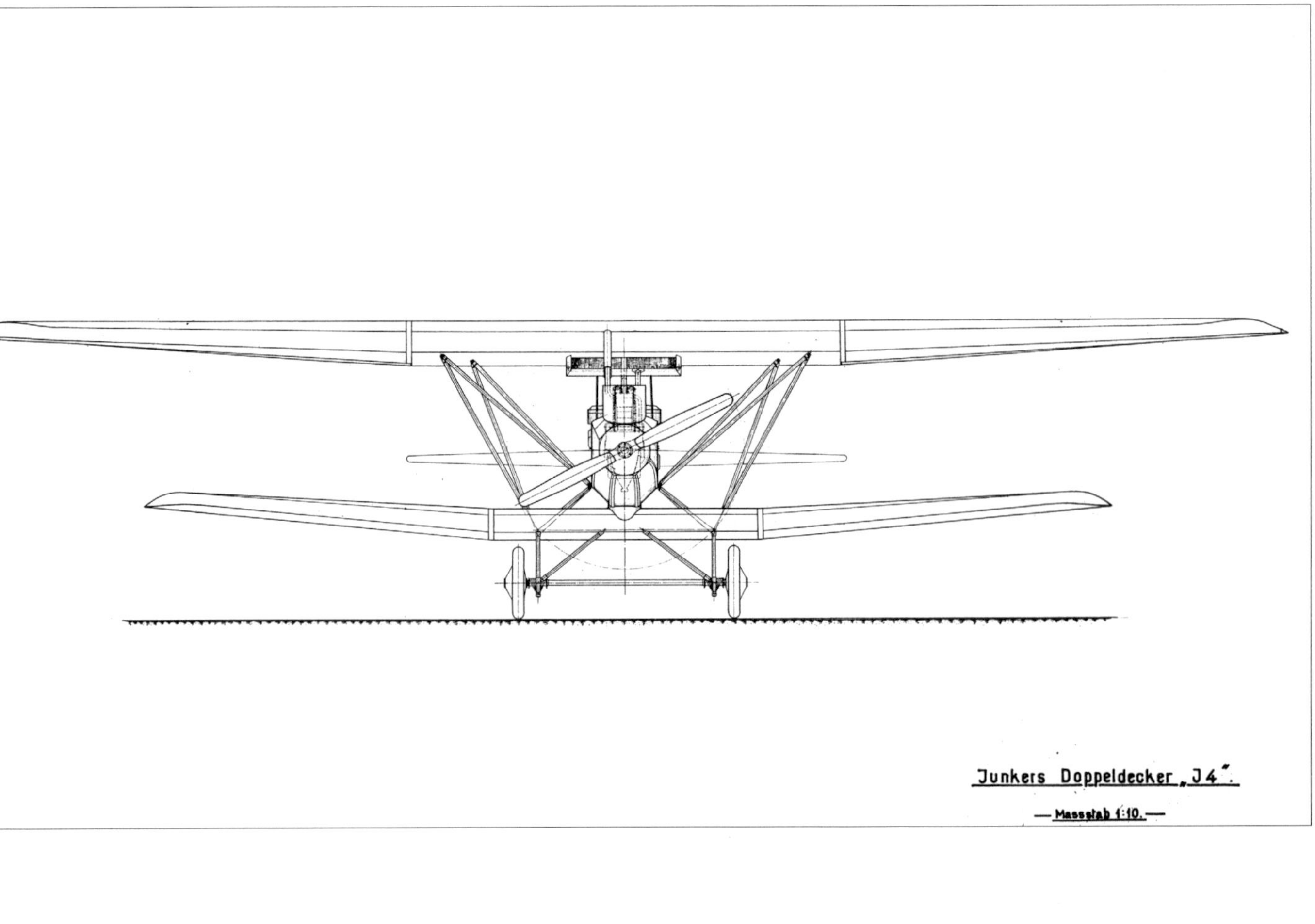
Junkers Doppeldecker „J4".
— Massstab 1:10. —

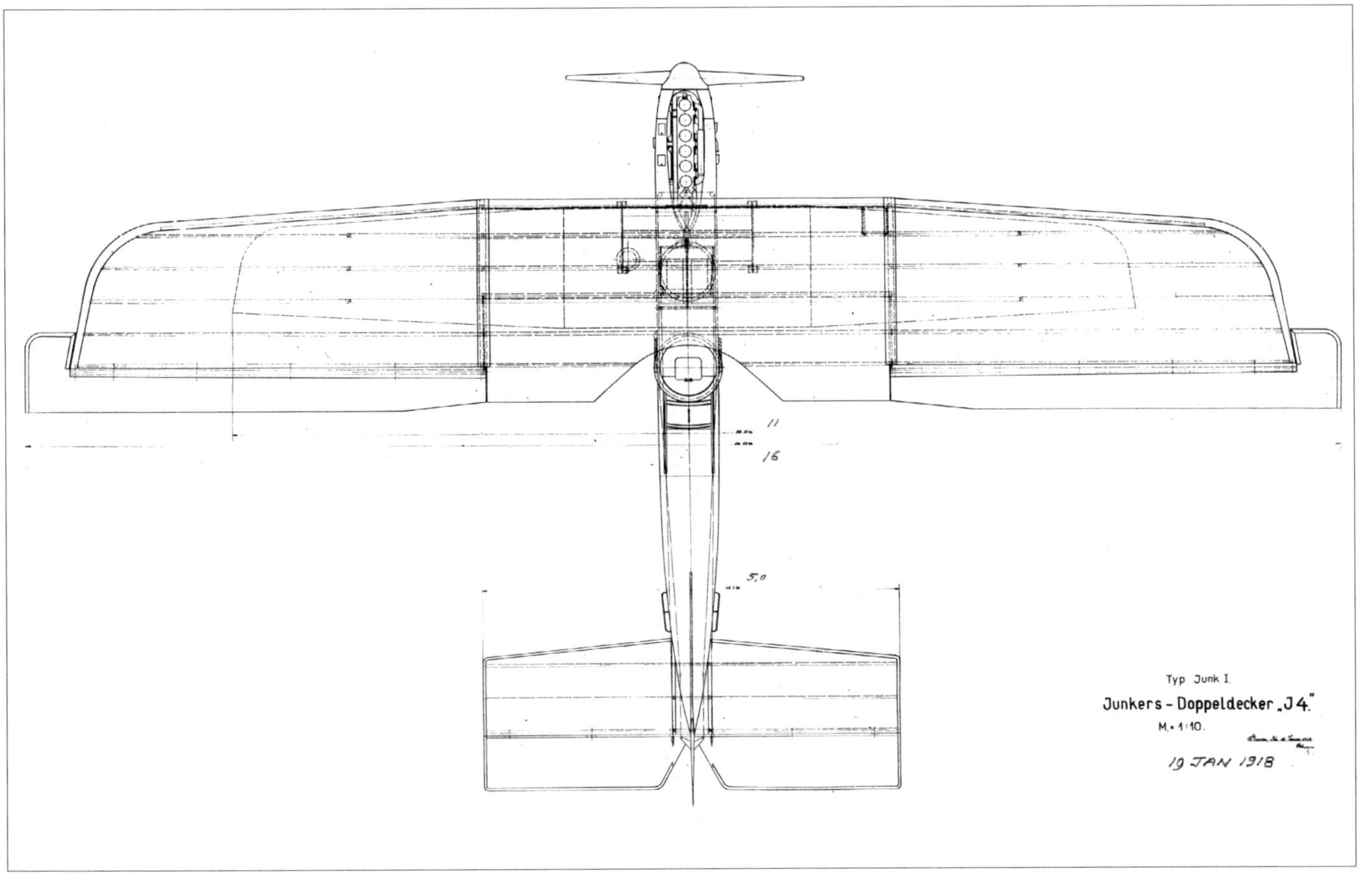

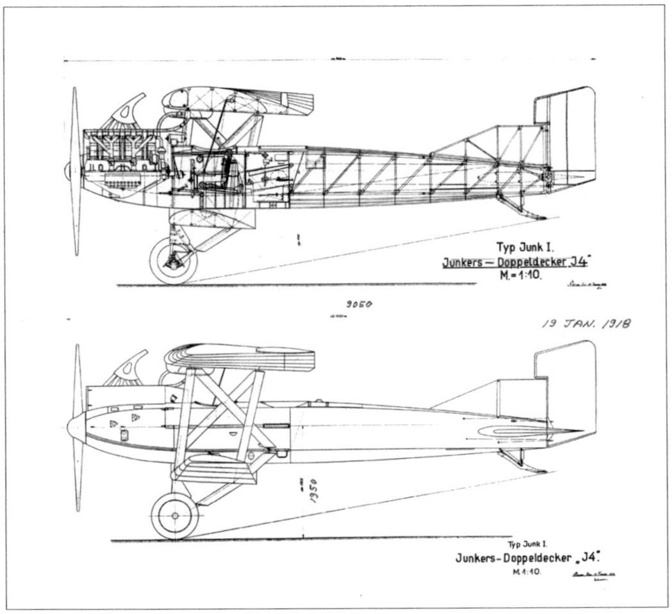

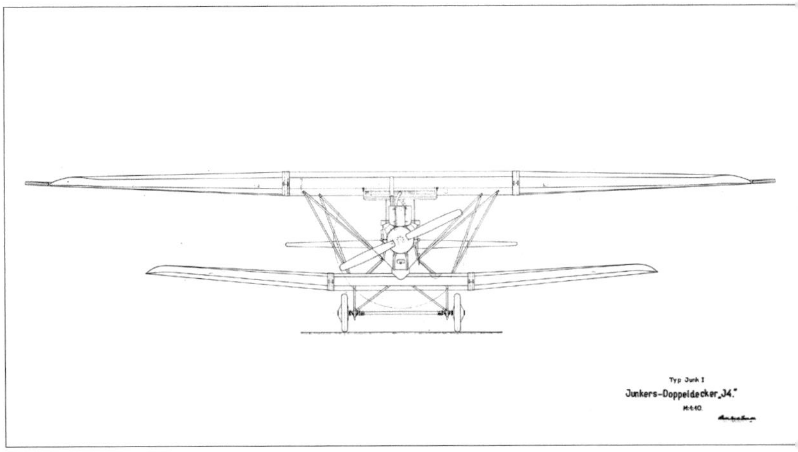

This Page: Dated Jan 1918, these drawings show the production J 4.

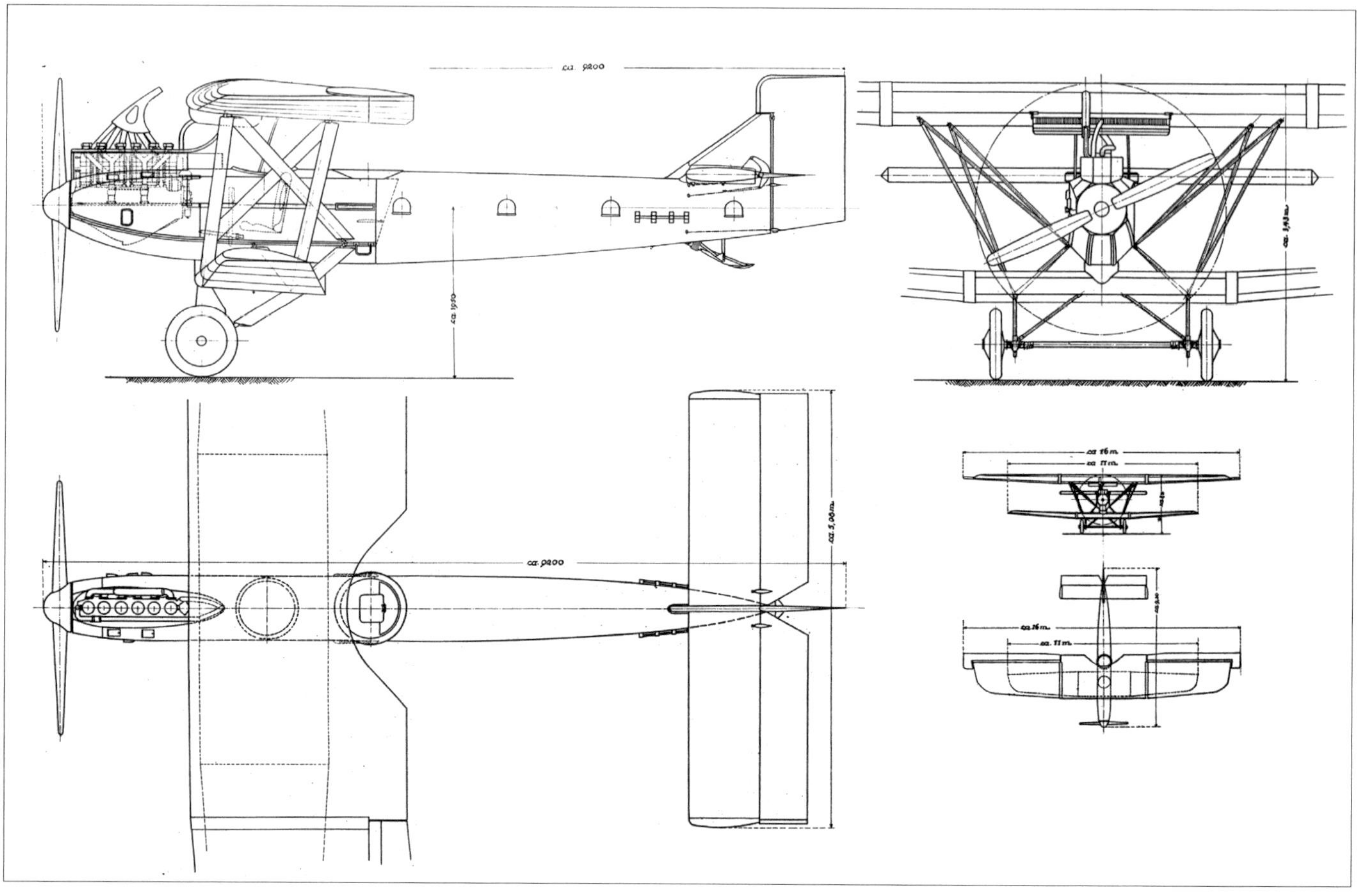

Above: Dimensioned drawing of the proposed J.II. Note the small fin and rudder.

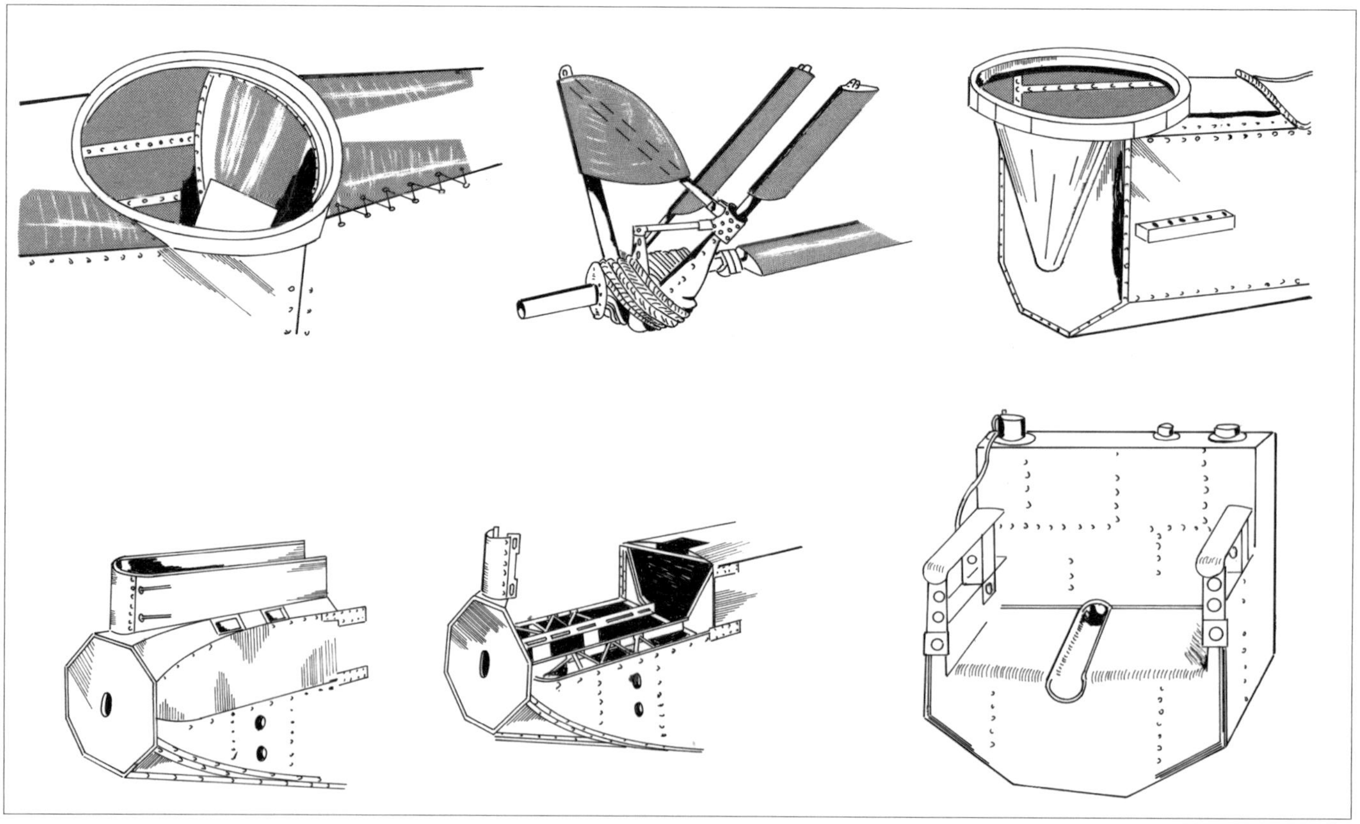

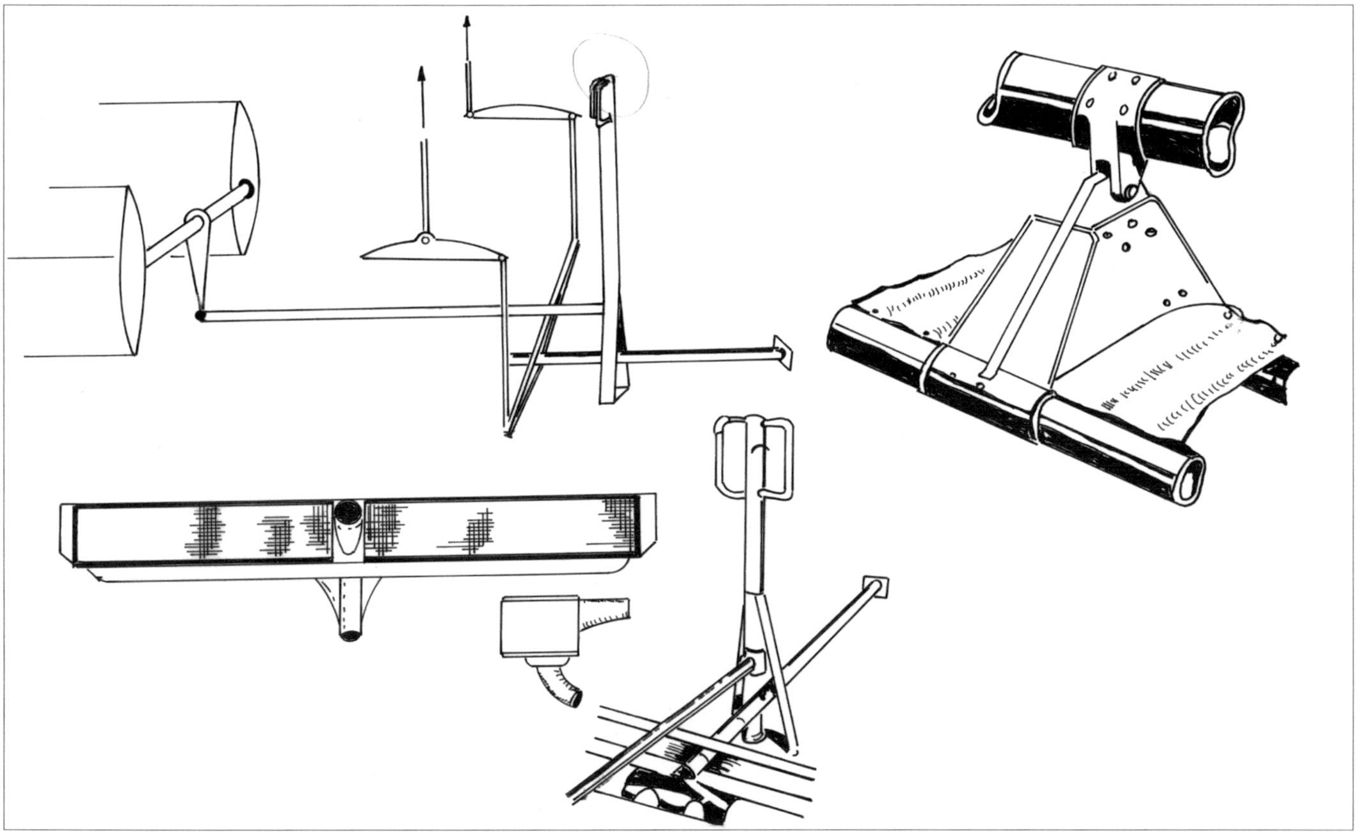

Above, Below, & Facing Page, Bottom: These sketches were reproduced in contemporary aviation magazines in 1919–1920. Some appear to have been incorporated into the USAS reports on the Junkers J.I. These have been re-drawn by the author.

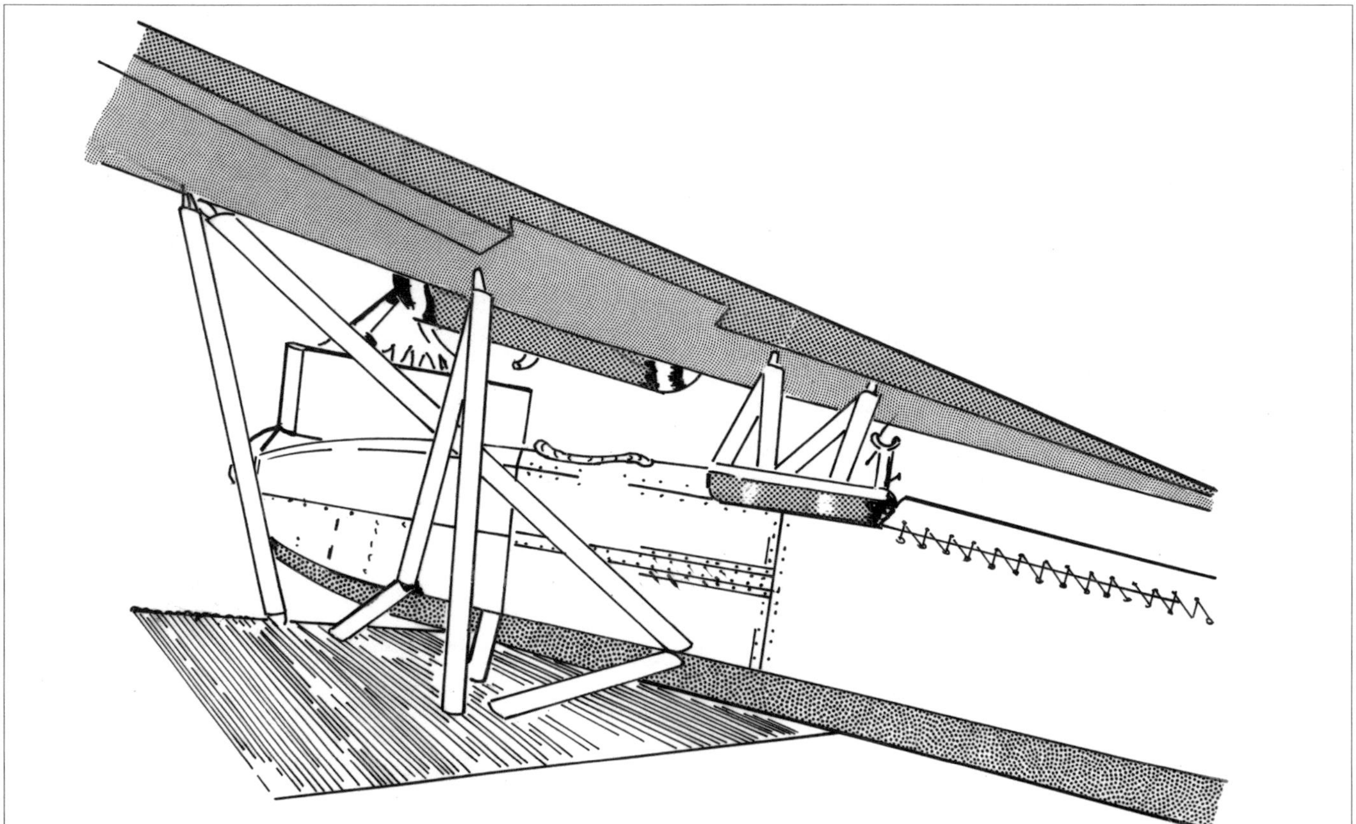

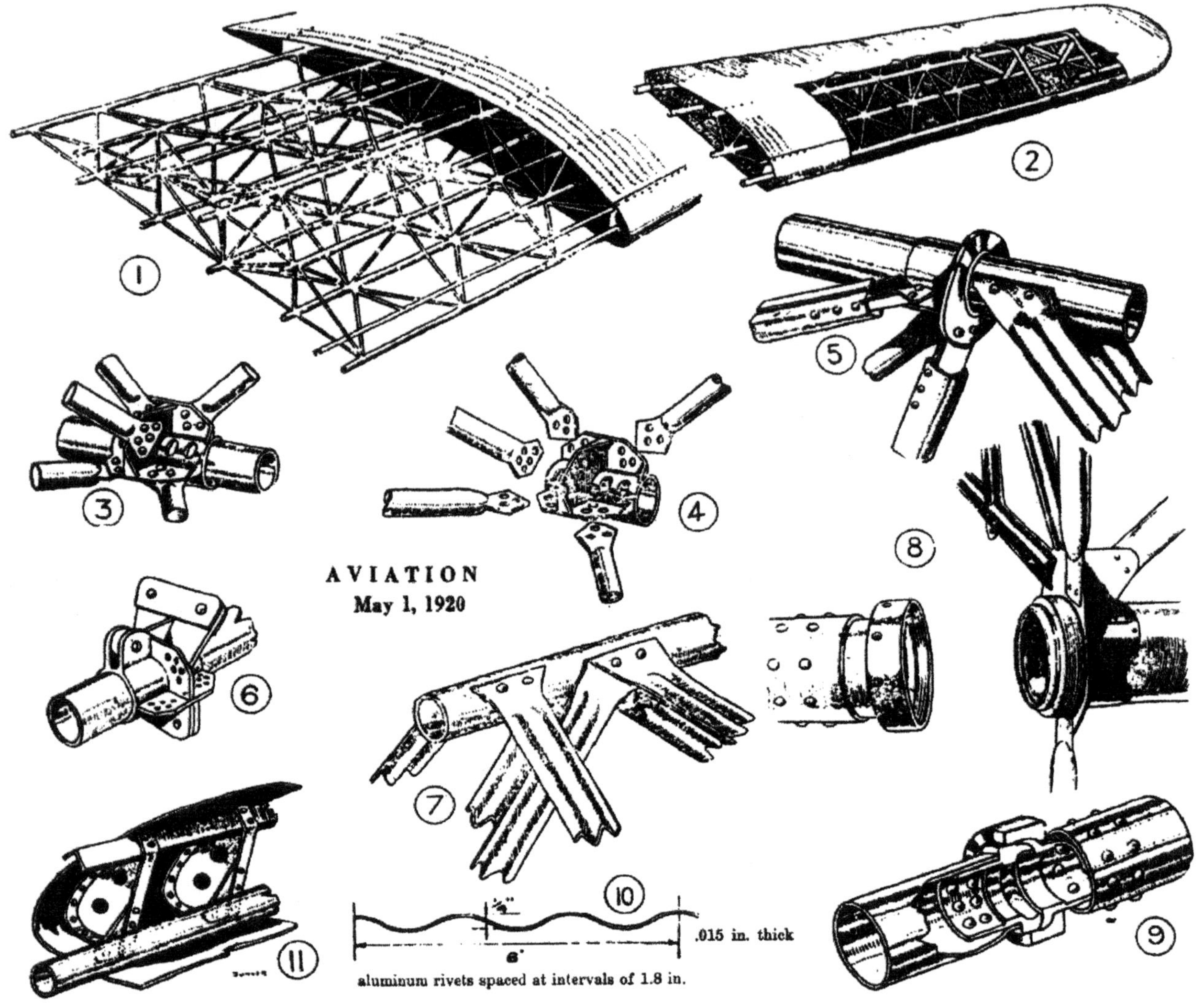

Constructional Details of the Junkers Wing: (1) and (2) General Construction; (3) to (7) Details of Wing Braces; (8) and (9) Details of Wing Attachments; (10) Corrugated Sheet Covering; (11) Leading Edge

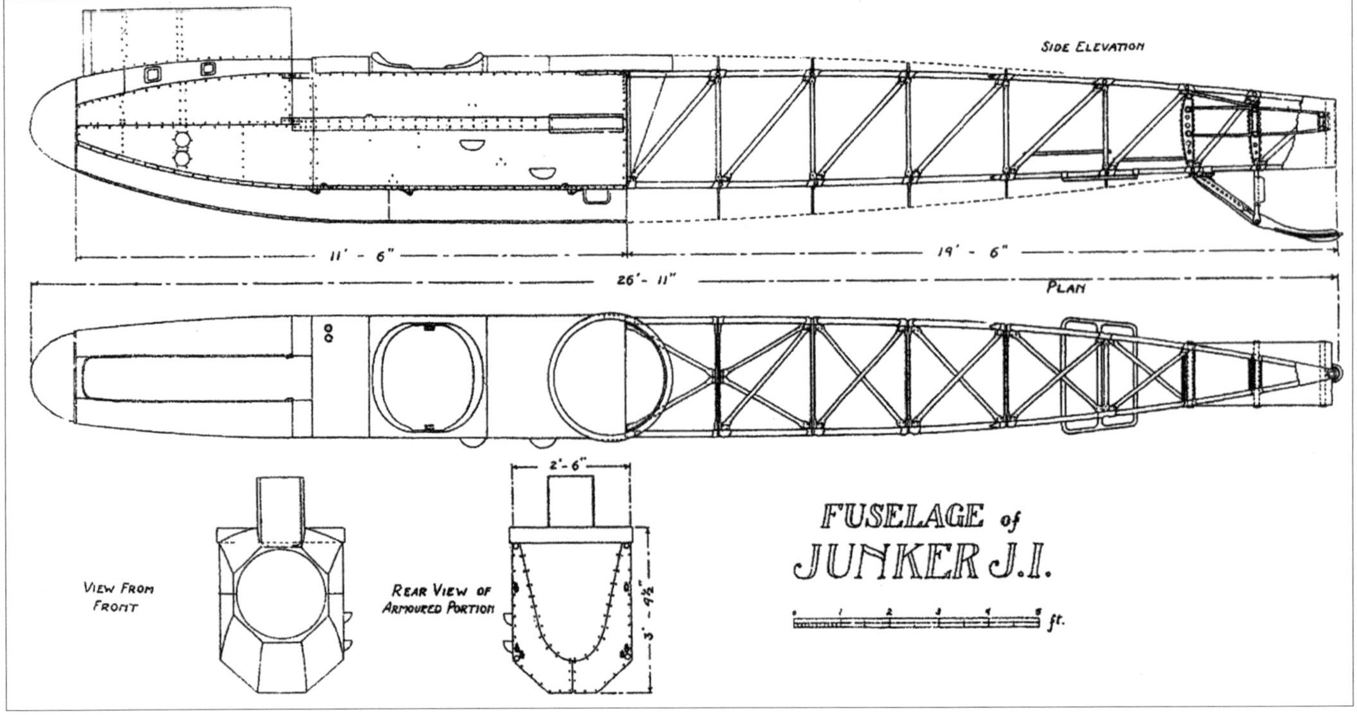

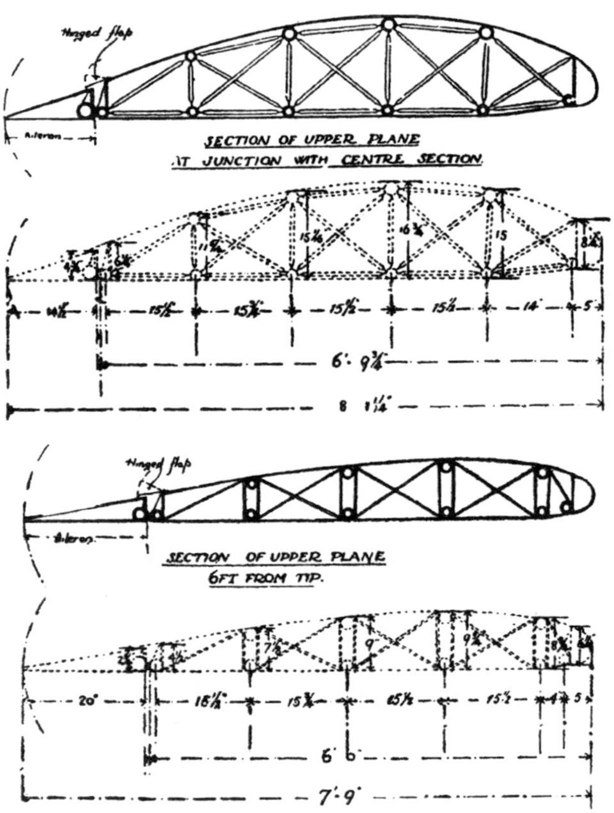

FIG. 3 JUNKERS WING SECTIONS

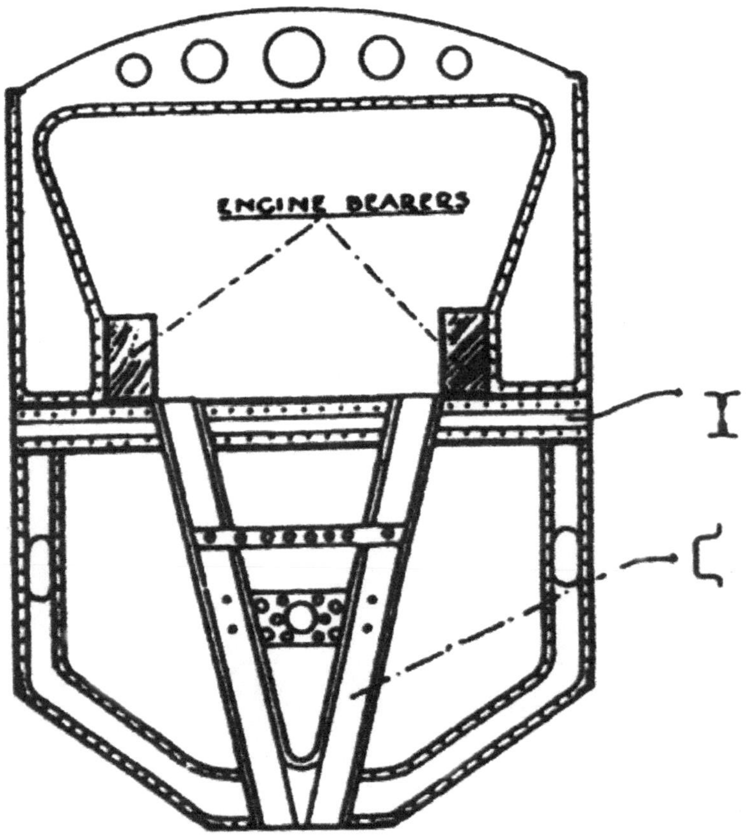

The engine installed is a 230 H.P. Benz, and it is remarkable that in such an all-metal type of machine the engine bearers are of ash. The bearers have sheet steel supports, and are braced to the side by means of duralumin tubes of 20 mm. diameter. Fig. 29 shows how this is done.

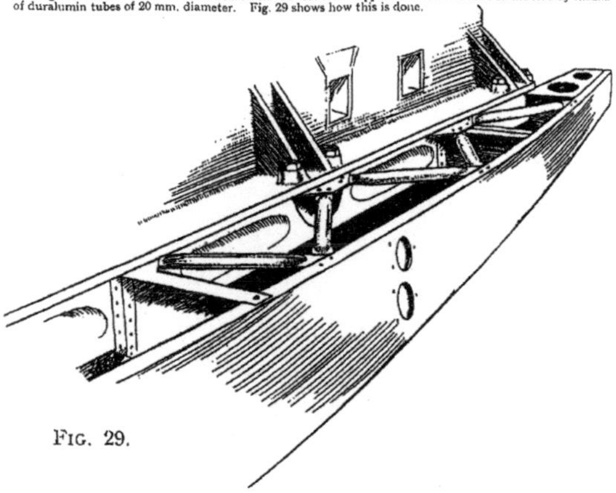

FIG. 29.

The Junker elevators are controlled by means of stranded cables attached to vertical levers, which are actuated by a steel tube direct from the control column. The extremities of the two vertical levers are not equidistant from the pivotal tube, but the upper part of each is appreciably greater than the corresponding lower part. It is found that the upper and lower portions of the elevator kingposts bear the same relation to one another, and the idea is evidently to reduce the mechanical strain on the upper control wire.

The rudder control embodies a curved steel tubular rudder bar, as shown in Fig. 25, and the customary cables and pulleys. Neither stirrups nor adjustment of position are provided. Two pairs of kingposts are fitted to the rudder post, and the cables are duplicated.

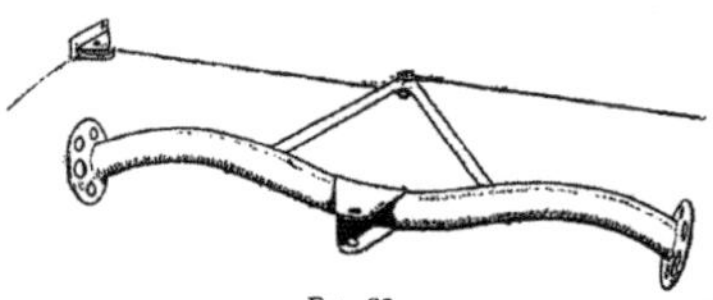

FIG. 25.

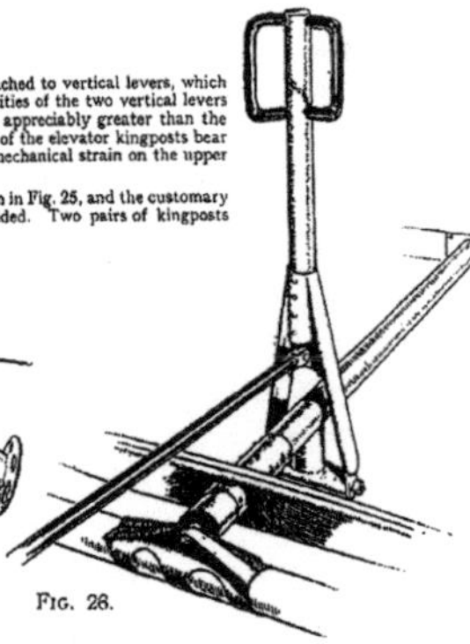

FIG. 28.

The junction of the two parts of the fuselage is very simple. From the outside, the longerons are seen to be butted up to the rear armour plate of the observer's cockpit, but no means of attachment is visible. Investigation from the interior of the cockpit shows that a substantial castellated steel nut is screwed over a steel bolt of about ½ in. diameter, and secured in place by the usual split pin. It was not possible to ascertain the exact manner of fixing the stud of the bolt into the end of the tubular longeron. Only the usual riveted steel collar to which the bracing tubes are riveted appears on the outside. The drawing (Fig. 20), shows the principle of this arrangement.

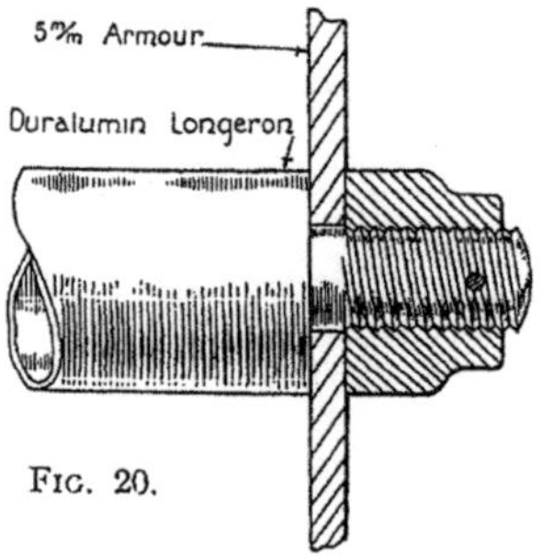

FIG. 20.

The construction of the divided elevator is very simple. At the front is a duralumin tube hinged by the means shown in Fig. 27 to the rear edge of the tail plane. There are six such hinges altogether—three to each half of the elevator. The corrugated duralumin covering is riveted to the spar and is also held together at its rear edge by rivets.

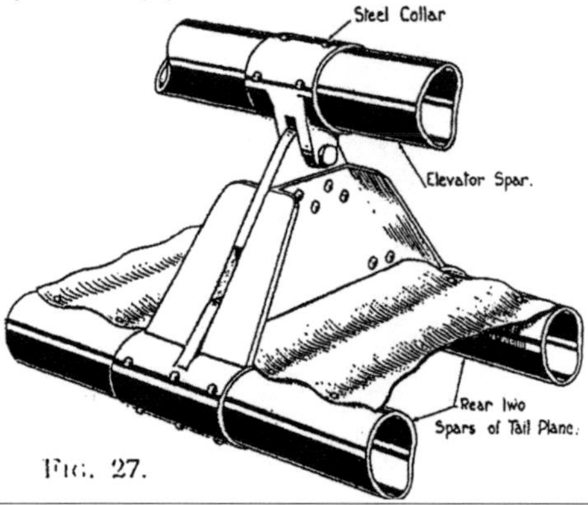

FIG. 27.

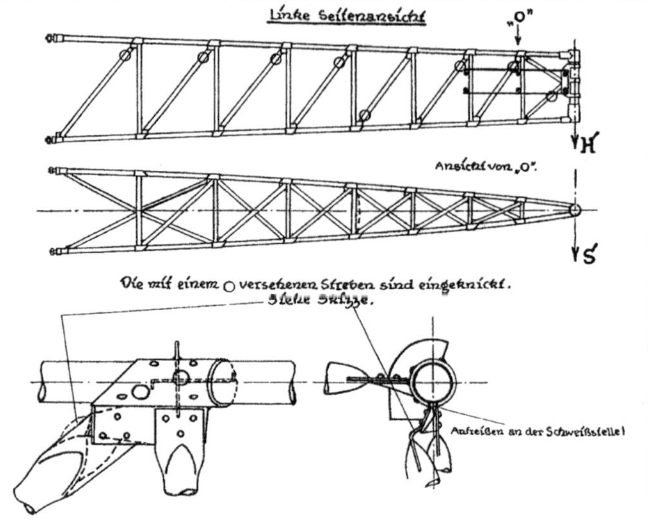

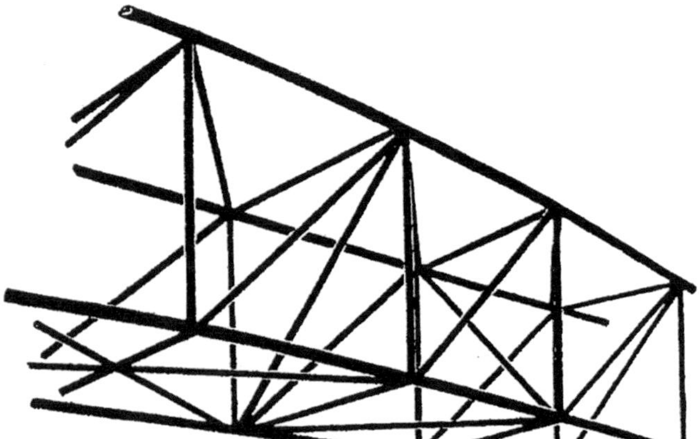

FIG. 18.

The rear portion of the body is built of duralumin tube throughout, and is covered with laced-on fabric. There are four longerons, so arranged that the fuselage ends in a vertical knife-edge about 16 in. long. Cross tubes are arranged horizontally and vertically at intervals. The diagonal bracing is not of wire, as is the case in the D.7 Fokker and the A.E.G. (which, it will be remembered, also have fuselages of metallic tubes). Diagonal tubes perform this function in the Junker. The arrangement of bracing tubes may be gathered from Fig. 18, and is worthy of careful attention.

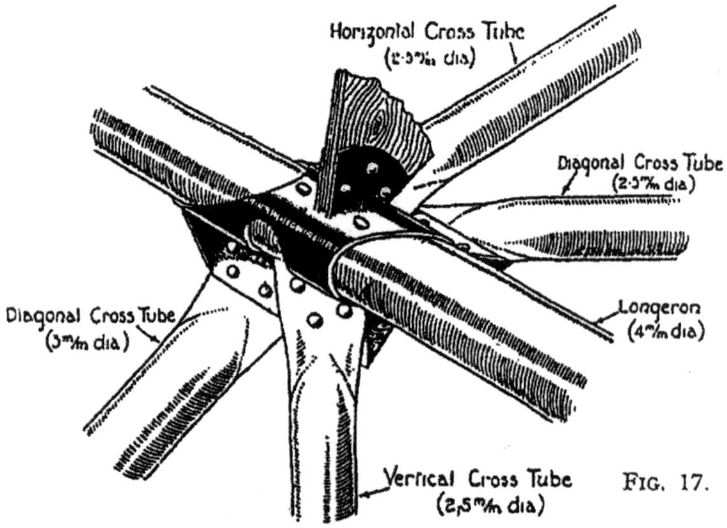

FIG. 17.

The junction of the duralumin tubes is effected by means of steel sleeves which embrace the longerons tightly and are pinned to them. The exact shape of one of these clips is shown in Fig. 17. It will be noticed that the cross and diagonal tubes are flattened at their extremities, and riveted to appropriate shelves welded on to the clip. Three-ply formers are fixed to the upper and lower cross tubes, and (in the case of the deeper lower formers, at any rate), are joined by light wood stringers which pass from the end of the armoured portion to the front end of the tail skid.

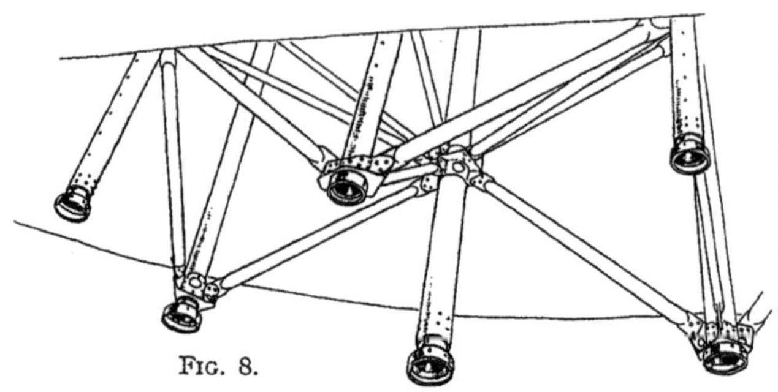

FIG. 8.

The constructional methods already described as being found in the upper planes are followed throughout both wings and centre sections except that towards the tips the place of the bracing tubes is taken by strips of duralumin longitudinally grooved to resist bending strains. Fig. 9 shows this, and also makes it clear that the steel collars to which the bracing tubes are riveted are not found where strips are employed. The finish of the tubular bracing and commencement of strip bracing is illustrated in Fig. 6.

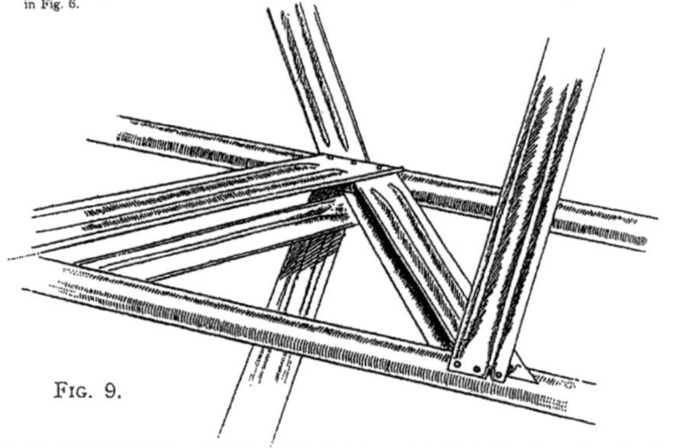

FIG. 9.

UNDERCARRIAGE.

This is built to form one unit with the lower centre section. There are two vees which are of normal design, except that a supplementary tube is welded to the upper extremity of the forward limb and to a point just below the middle of the rear tube (*see* Fig. 22). The vees are parallel and vertical. The shock absorbing device—of which a sketch may be seen in Fig. 22—does not present any unusual feature. Twin coil spring is employed and a double loop of cable limits the upward travel of the axle, while a radius rod controls the movements of the axle. It will be noticed also from Fig. 33 that each vee is stayed by means of a steel tube which connects the lower part of the front limb to the middle of the lower centre section, and that wire bracing is entirely absent.

The axle is a steel tube nearly 9 ft. long. In common with all other exposed tubes in the machine, it is faired with sheet aluminium, which is riveted to transverse webs (also of aluminium), which are in turn riveted to the axle tube. The fairing is therefore of the type that rises with the axle when the machine lands. There are no compression tubes fitted parallel to the axle, since the extra pair of tubes already mentioned render this unnecessary. Incidentally, they also eliminate wire-bracing. On other German aeroplanes which avoid wire wing-bracing (such as the D.7 Fokker and D.15 Pfalz) the undercarriage wires constitute the only external bracing.

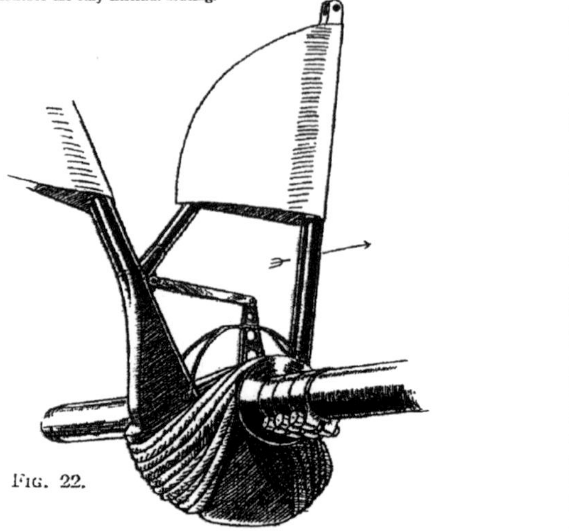

FIG. 22.

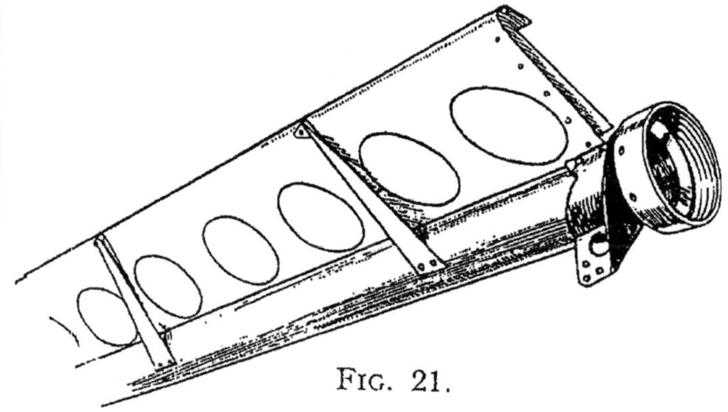

FIG. 21.

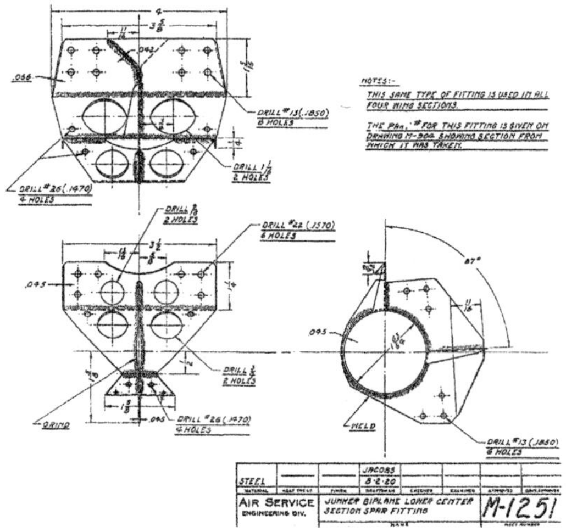

WING CONSTRUCTION.

It is in this particular that the greatest departure from established practice is found. The planes are constructed entirely of metal, even to the covering. A description of the constructional features of one of the upper planes will explain the system upon which the wings are built.

Each upper plane contains 10 spars of duralumin tube, without counting the aileron tube. Page 3 gives a section drawn to scale, showing the disposition of these spars, and the lower half of the diagram gives all dimensions, including the diameters and gauges of the tubes. The tubular spars are braced to each other by means of smaller tubes, also of duralumin. The exact system employed is shown very clearly in the photograph (Fig. 7) and the sketch (Fig. 8). It will be noticed that steel sleeves are riveted at intervals along the spars, and that the bracing tubes are flattened out at the ends and riveted to small flat plates welded on to the steel collars. Examples of these joints are given in Figs. 4 and 5.

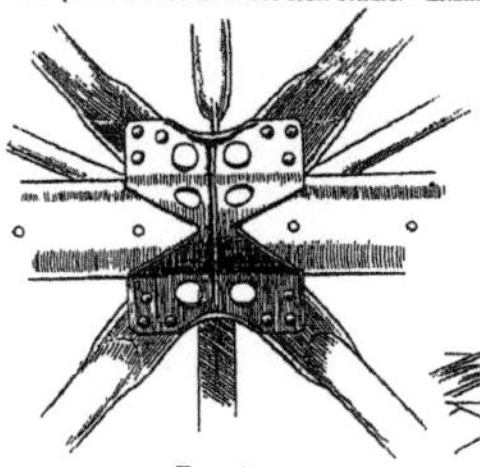

FIG. 4.

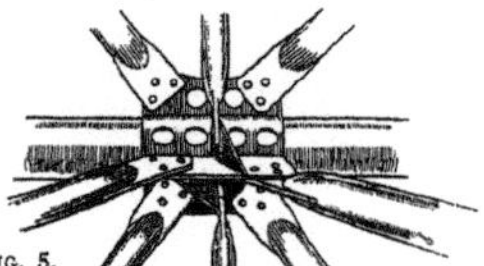

FIG. 5.

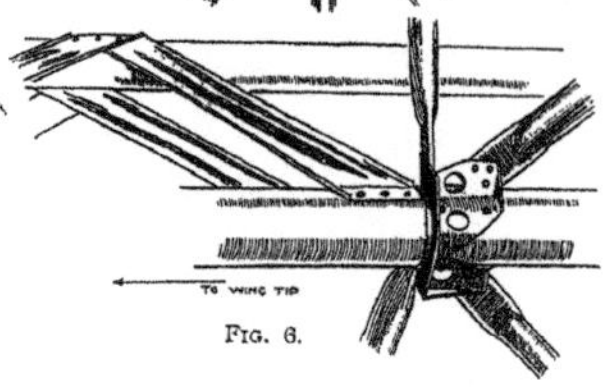

FIG. 6.

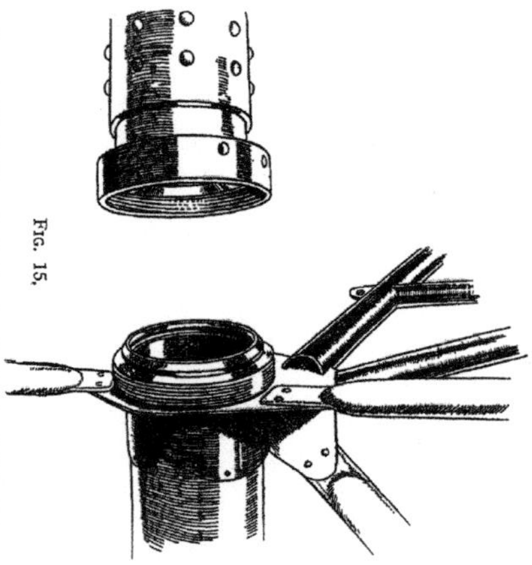

FIG. 15.

The way in which the wings are joined to the centre sections is simple and effective. Reference to Figs. 15 and 16 (which are not necessarily drawn to scale), shows that each of the spars is fitted with a steel sleeve which fits inside the duralumin tube and is riveted in place. One sleeve carries a threaded collar bevelled as shown. The opposite spar has a similar internal liner of steel, riveted in place, and a loose internally threaded steel collar. The end of the liner is bevelled to take the bevel of the opposite spar. Thus, when the bevels are fitted together, and the collar screwed on to the male thread (tommy holes can be seen in the sketch), a firm and rigid joint is made. When it is remembered that all the numerous tubes are joined in this manner, it is evident that the junction of wing to centre section is of great strength. Indeed, the designer has trusted solely to these joints to take all lift, drag, and landing strains, for there are no other attachments of any sort between wings and centre sections.

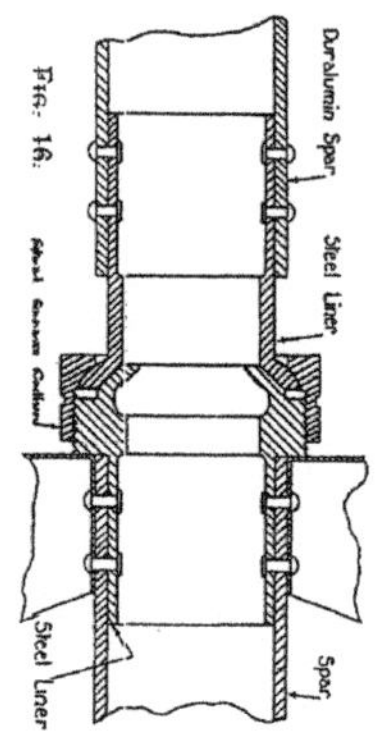

FIG. 16.

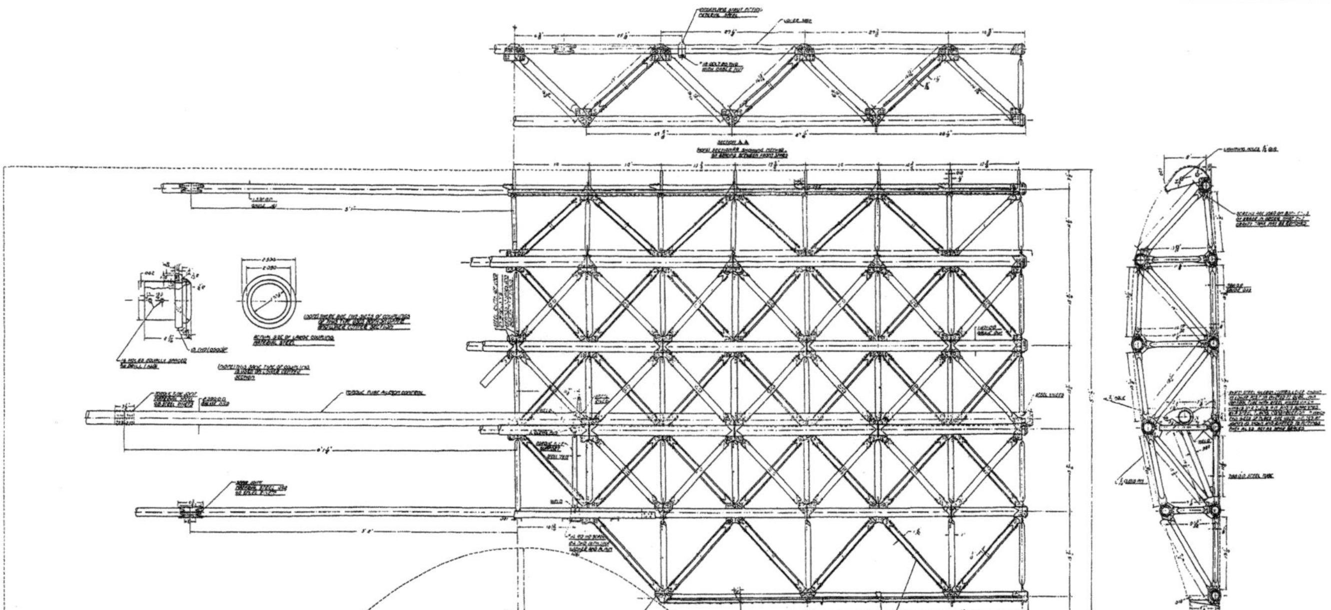

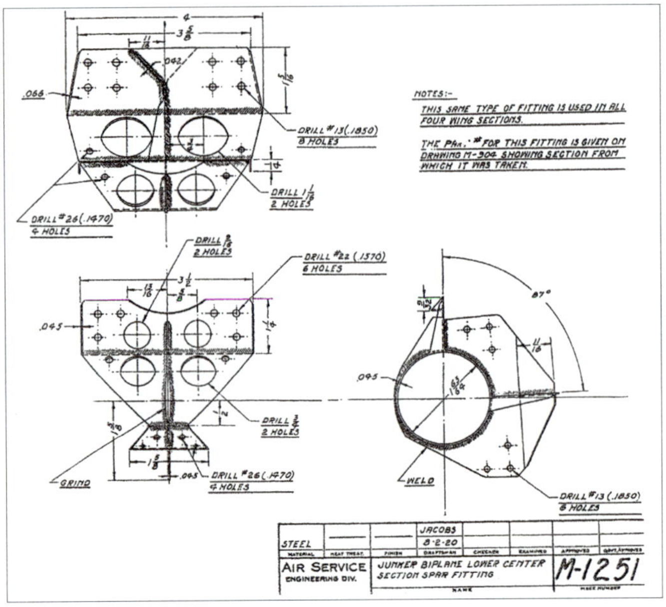

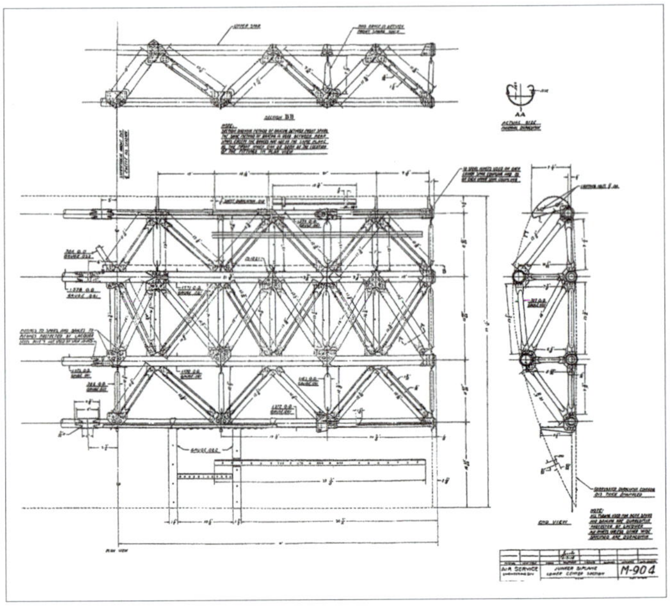

This Page: The USAS was very interested in the Junkers J.I and did a detailed examination of the type. The following drawings were produced from these. Unfortunately a full set of drawings and reports has not been found to date.

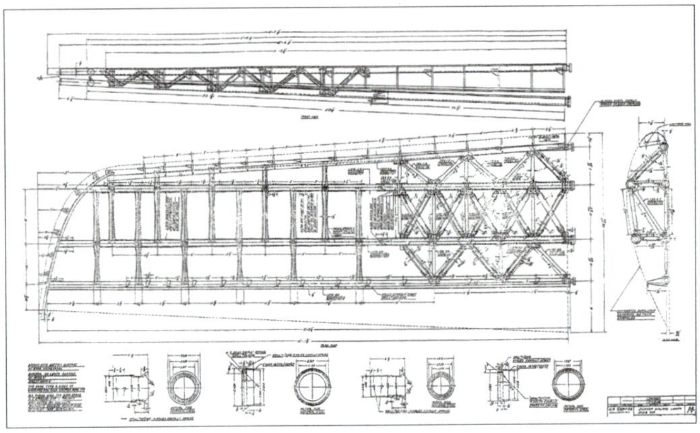

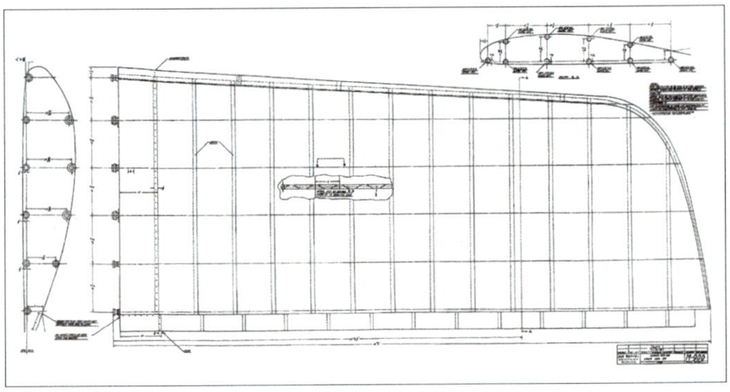

19 and 250.

Both speak unanimously and positively (about the J.I). However the views of the pilots and observers differ markedly. The machine is described as "Good" by *Fl Abt (A)* 250, however the overhead view was considered unsatisfactory by *Fl Abt* 19.

1. About the Armoured Biplane Type Junkers 1

The J-I-F has distinguished itself due to its good flight characteristics and the greatly enhanced safety from damage from bullets or fire from machine guns or artillery.

Even though the takeoff requires a longer run, which is natural for such heavy aircraft, it surprises one with its short landing compared to other armoured aircraft, for example the A.E.G.

The aircraft is relatively stable even in gusty conditions. Only the strongest gusts require correction. The view of the pilot and observer is unrestricted and is valued. The armour has proven to be effective enough. The radiator which is situated under the upper wing was damaged by splinters from machinegun bullets bursting on the fuselage armour. Critically this splinter damage led to a forced landing and could be avoided by a radiator slightly narrower than the fuselage width. The question is would such smaller radiator be sufficient. The current radiator is sufficient. During the emergency landing in our artillery positions the aircraft hit a shell-hole and somersaulted. The centre section of the upper wing was heavily damaged. At both strut mounts the *langerohr* (literally the long tube) was ripped out and all struts were dislocated from this point and bent. During salvage operations the aircraft received further damage to the right rear elevator and skin by artillery shrapnel.

The aircraft cannot be made airworthy by the unit.

The *Abteilung* extensively tested the armoured A.E.G. aircraft as well as the Junkers armoured biplane at the front. Both are preferred by *Infanterie-Flugzeug* flights over non-armoured aircraft. The

Junkers J.4 (Military J.I) Serial J.586/18
Photo Essay
by
John Weatherseed, The Vintage Aviation Team

Above: View of the upper wing root at the centre section showing the nine (9) tubular spars that the structure was built around.

Right: Upper wings in storage. Despite the many injunctions in the manual for the type stating that the wings are not to be stored on their edges, this is the most convenient way of storage and hence was carried out from the first when the J.4 biplanes were taken into Allied hands after the Armistice. Points to note are the corrugated cover over the aileron gap and the sand-coloured lower surfaces.

Above & Below: Close-up on the fuel tank in the upper wing centre section. Note the maker's plate.

Above: Close up of fuel tank maker's plate.

Right & Above Right: View from rear showing straps holding fuel tank in place in upper wing centre section.

Above: Close-up of ZAK mark on fuel tank.

Right & Above Right: Close-up of rear section of the upper centre section wing root.

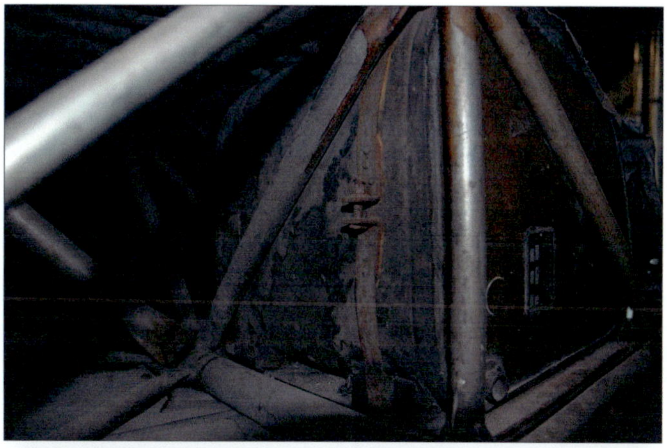

Above: The connection for the wing sections was arranged such that the machine was correctly aligned when assembled and no rigging was necessary. Note the riveted connections on the flattened tubes.

Above: Plate on bottom of upper wing reads: Abnahmetag (date of acceptance) 3 9 1918–Geprüft (Checked) Nr. O.M.1; With the ZAK and crown at the bottom. Note the sand coloured paint.

Junkers armoured biplane however possesses many advantages over the A.E.G. Both types require very experienced pilots.
Signed: Minarski[29]
Hauptman und Abteilungsfuher (Captain and Squadron leader)

2. *Fl Abt* No. 19. Report on the Junkers J aircraft.

B.Nr.(Bericht Nummer/Report number) 2755
o.U the 28.09.17
To *Gruppenfüher* (Group leader) *der Flieger* 15 (of

Facing Page, Lower Left: View of the centre section bell-crank for the aileron.

Above & Below, Left: Steel connections for the interplane struts to the centre section. Note the green of the strut streamline covers.

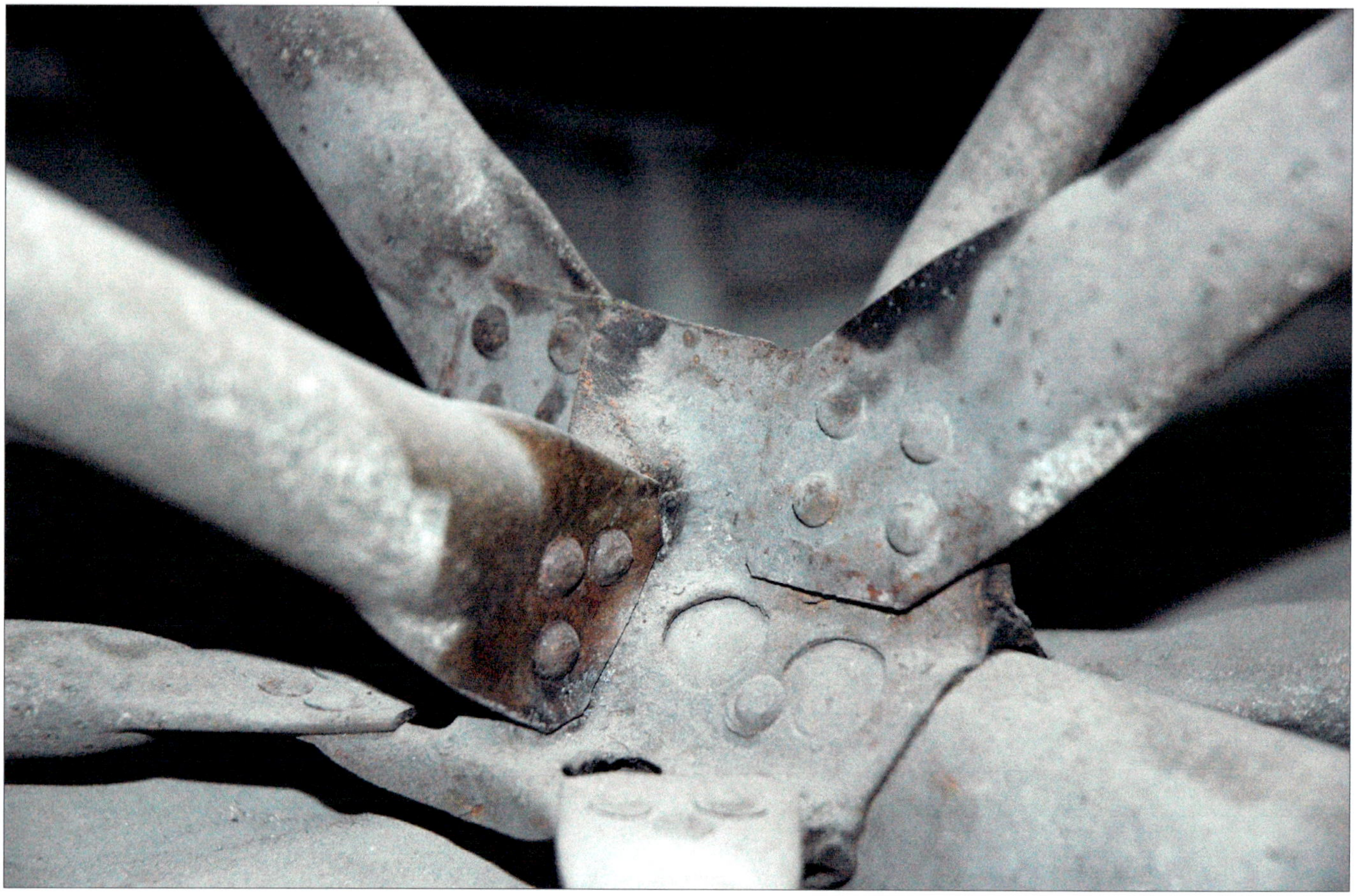

Above: A join within the wing showing the flattened ends to the duralumin tubes and the riveting of the ends of the tubes to the connector that in turn is attached to a tubular spar.

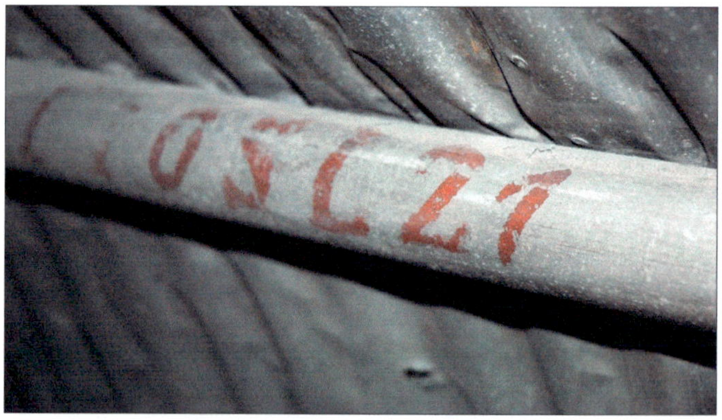

Above: Another marking on a spar.

Left: One of the markings on the wing spars. Assumed to be "J" for Junkers.

Above: Crudely painted marking on this spar.

Above: Note the shaped steel piece that is attached to the spar and the way it is formed to take the duralumin tubes that form the internal structure of the wings.

Above: Trailing edge, upper starboard wing at aileron gap cover flap.

Above: The pitch of the corrugations was 50 mm. The ends of the corrugations leave a smooth transition to the front of the sheet for the facing piece to cover and preserve the aerofoil shape.

Above: Upper starboard wing aileron. Note the aileron gap cover cut from strips of corrugated metal.

Above: Stencil on bottom surface of upper wing. "Junk I 586".

flight 15) with a request to further distribute.

The Junkers J aircraft is quite suitable for the purpose it was built for but at the same time the

Above, Top, & Below: Front edge of the aileron with capping strip riveted to the structure. Rivets are used in all cases. Note the way the corrugations are finished at the leading edge.

thought occurs that armoured *Infanterie-Flugzeug* aircraft are capable of many improvements. In regard to the handling qualities the unit has made the following observations:

All section pilots liked to fly the machine as it caused them very few difficulties. Even relatively inexperienced pilots were able to fly the machine. The biggest disadvantage in a flying technical sense

Above: The front edge strip on the upper wing. Note how the corrugations are bent to conform to the curved shape and how the edge strip overlaps.

Above: A C-channel section inside the wing.

Above: The riveted C-Channel attachment.

is the take-off. This is unavoidable considering the massive weight of the machine but acceptable because of the huge advantages the machine has. In the air the machine is relatively manoeuvrable and easy to fly despite its great weight. The speed is satisfactory for an Infantry Aircraft.

Application of throttle and diving rapidly accelerates the machine. This is very comforting considering the relatively weak armament. The machine is very stable in the turn. She has a high glide-ratio. Because of this it is often possible after engine failure even at low altitude to make it back over the front (lines).

Apart from that it makes landing easy.

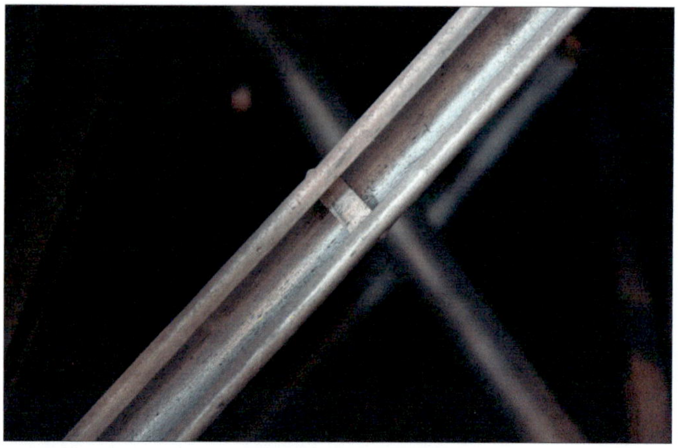

Above: Detail of a C-Channel.

Above: View inside wing showing tubes and C-Channel sections.

Above: View of the front of the wing from inside. The shaped duralumin piece that gives the front curvature to the aerofoil is clearly shown.

Above: Detail of duralumin frontal piece. Note stay.

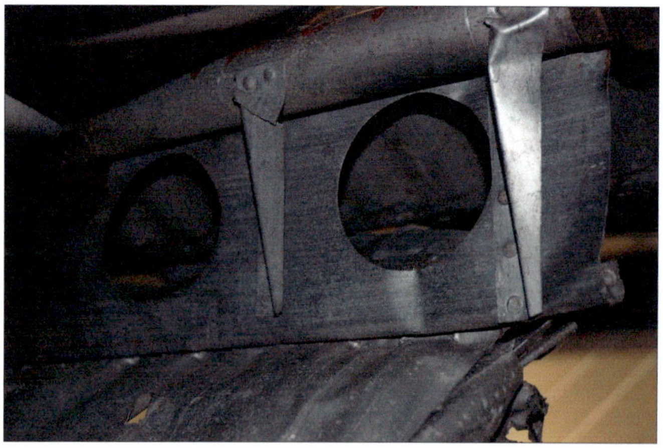

Above: This appears to be a stiffener behind the leading edge.

Above: The upper wing still bears the original green and purple camouflage.

Above: The 2-1-5 annotation meaning has been lost but is most likely an early accession number. Variations of this numbering appear on many of the German War Trophy items in the Canadian collection.

Right: Stencil on the armour tower surrounding the upper part of the engine.

Below Right: Detail landing gear cords.

Below: View showing the filler cap for the radiator on the upper wing centre section. The cap is on the centre-line of the wing.

Above & Above Right: The corrugated skin has suffered over the years, particularly the leading edge of the wings.

Above: The clips holding the lifting handles to the rear fuselage are evident in this view.

Above: Front view of the machine shows the damage to the wing leading edges.

Above: Three-quarter rear view shows remnants of the original camouflage pattern.

Above & Facing Page, Top: Fin and tail skid attachments, and rear fuselage detail.

Left, Above, & Right: Details of the gunner's cockpit. At the right side is a wooden plate marked *"Platz für Schaltkasten"* ("Space for switch box"). At front panel with the mount for Telefunken transmitter, near typical electrical socket, above the rectangular aluminum control cable cover.

Additional photographs of J.586/18 by Garry Barling (This Page & Facing Page)

Above: Strut to fuselage detail.

Top: Close-up of the engine installation.

Right: The access hatch to the engine is missing.

Above: The J.4 had good aerodynamic lines.

Right: The antenna hole in the bottom of J.586/18.

Below Right: Detail The panel along the edge of the cylinder housing is not armoured.

Below: Footstep detail.

Junkers J.I Fuselage in Milan Museum

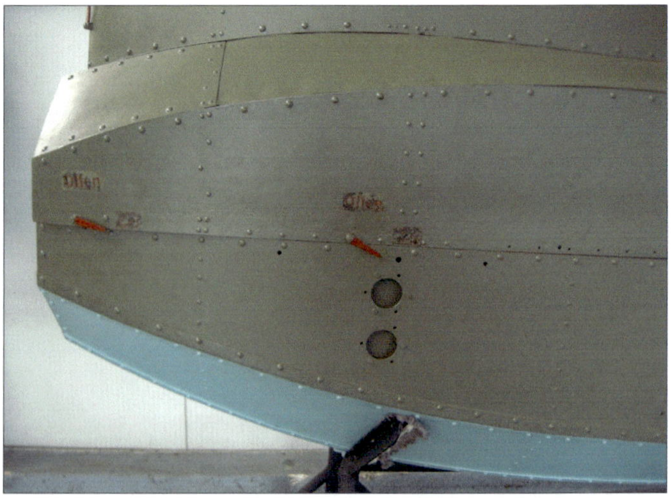

This Page: The Junkers J.I fuselage in the Milan Museum after restoration. Note the un-restored "windows" with preserved original paint and stencils. The side flare boxes are missing.

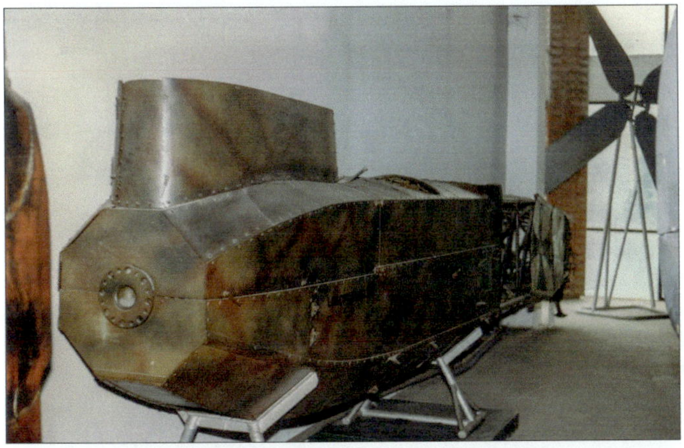

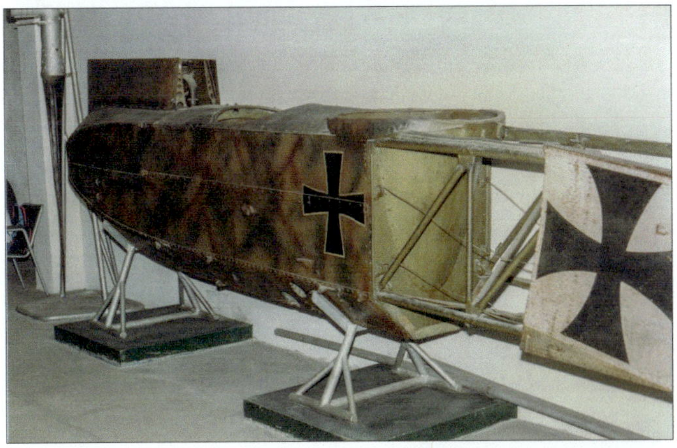

This Page: Details of the Junkers J.I fuselage in the Milan Museum. The upper left photo shows it after restoration and the other photos show it with its inaccurate camouflage before restoration.

This Page: Junkers J.I fuselage in the Milan Museum before restoration.

The cockpit layout for the pilot as well as the observer is good and easily scanned. The infantry aircraft often flies at low altitude and in strong gusty conditions. With this heavy machine the pressure generated on the ailerons is so great that the pilot does not have sufficient strength to operate the control stick. Wheel control is therefore urgently required. That is the request of all the unit's pilots.

The machine is mainly suited for scouting/observing troop concentrations in the rear area and support trenches up to several kilometres behind the front. To determine the position of our own front line we do not use armoured aircraft. In flights to establish the location of reserve forces behind enemy lines the J 1 is forced to fly below 100 meters. The following is said about enemy defences: - On the big battle day of 20.09.17, it was revealed that the enemy was working to a plan to complete his ground based aerial defence with machine guns. Proof of this are the increasing number of M.G. hits on these flights from the ever increasing ground based M.G. fire in the enemy's rear areas. Against this kind of enemy ground based fire the machine is safe and absolutely reliable. She shows her absolute solid construction despite hits in the primary structure. She returned from an infantry flight with twelve hits, one in the undercarriage brace and survived a strong bounce on landing. Once she returned with a broken strut and numerous hits in the wing. The armour had five hits which were only discernible by the paint flaking and minor indentations. Because of these facts the crew enjoy a great feeling of safety and, as a result, this enables calm and workmanlike observation. The effect of enemy flak in our experience is not a problem. Flak fire is strongest at heights above 4-500meters, below 200 meters nowadays one is never fired upon by flak.

The attack by enemy aircraft is the most uncomfortable situation in a Junkers Machine. Here above all else there have to be improvements. Because the J-F-I always flies low the enemy is invariably above it. The machine is unsuitable for aerial combat with the enemy from above. Even though relative to its weight she is quite manoeuvrable she will always be at a disadvantage to a C-Type aircraft or a fighter. The following must be considered. During the last big battles the enemy frequently employed blocking squadrons at low altitude to deny our infantry aircraft access to their rear areas. More defensive weapons cannot overcome this problem. The infantry aircraft will always require a fighter escort. It is still troubling that the view from the Junkers aircraft is especially poor overhead thus making it easier for the enemy to surprise us. Also, the fighting ability could be greatly increased if the M.G. ring could be modified to allow the Observer to fire forward above the wing.

The pilot of the machine has almost no view above. He can see only forwards but hardly at all upwards. In any case the amount of armour could be greatly reduced if the armoured area were better utilised by concentrating the critical parts. This would undoubtedly cause problems as the forward movement of the observer will restrict his field of vision.

Because the observer is seated so far behind the pilot, communication (between the two) is difficult. The necessity to armour the radiator engenders mixed views. Complete protection of the radiator is likely impossible. However, the machine with a shot up radiator is still flyable for a short while and with the current installation this would appear to be a rare event. The radiator of this machine has not been hit. It would be a sure benefit if the 200-hp Benz was replaced with a 260-hp Mercedes. The weight increase would be slight. The long takeoff run would be shortened and the low level speed increased. High speed when overflying enemy rear areas is the best protection against enemy aircraft. The ability to observe would not suffer from this increase in weight.

The metal covering of the struts has proven unsuitable as a material. The wood covering as in the Rumpler C.IV would prove better. As the infantry aircraft is more exposed to infantry and M.G. fire than any other it would be worthwhile to protect the propeller splintering from bullet hits. The weak outer 40-50cm could be covered in aluminium or fabric. Fabric appears to be preferable.

Repairs can be easily done using the supplied special Junkers tools. As the section had no suitable tent the aircraft could not be housed. The weather had absolutely no effect on the machine. The paint on the machine has proven to be good camouflage. Even our own escort aircraft had trouble maintaining the machine in constant sight. It is recommended that the size of the markings be increased as the machine closely resembles the enemies R.E. (*presumably a reference to the R.E.8 with its long overhang of the upper wing*). It would also be beneficial due to the smaller lower wings to paint the markings on the underside of the upper wing. *(Reference to the photographs will show some J.I biplanes had the national markings on the lower surface of the upper wing.)*

It is foreseeable that in the future J-F-I aircraft will be needed even more than now.

I.V.
Signed. Leon
Oberleutnant.

The Junkers J.I in Combat

Ltn Wilhelm Paul Schreiber flew Junkers J.I 134/17 with *Flieger Abteilug (A)* 221. The unit had moved in great secrecy to the aerodrome at Was au Bac, ten kilometres north of Cambrai, as part of the preparations for Operation Michael. On 15 March 1918, Schreiber and his pilot *Feldwebel* Ernst Schäfer undertook an infantry cooperation mission when their Albatros J.I 705/17 was shot through the radiator forcing them to land amongst the shell holes, fortunately within German lines. The crew were unhurt but the machine was damaged.

On 21 March, the day of the big German breakthrough, the pair carried out more missions in J.134/17. It is known that on their first flight that day they plotted the extent of the German forward positions and confirmed that the enemy occupied the town of Demicourt. Two days later they reported that they flew at altitudes of 10 to 300 metres in order to ascertain where the troop positions were. Although they were hit on many occasions by ground fire they were able to continue their mission and present a detailed report.

On 27 March the unit moved eight kilometres east of Douai, to the airfield at Masny. These moves were always dangerous for the heavy Junkers sesquiplanes that required a compact runway, a condition that was not always available under such conditions as this advance. From available records it is known that Schreiber and Schäfer received an additional 66 new bullet holes in their aircraft on the 28th. On 6 April the pair was recorded as having encountered a Sopwith fighter while on a photographic reconnaissance flight. *Fl Abt* (A) 253 used a variety of aircraft and it is not sure if they were flying the J.I on this occasion.

On 5 May the unit moved to Abele airfield. The pair carried out dangerous infantry cooperation patrols nearly every day in J.I 134/17. Schreiber claimed a French Spad fighter on the 15th and confirmed this victory by taking a photograph of the crashed aircraft with his aerial camera before leaving the site. Their luck ran out on 30th when Schäfer was shot in the head and their Junkers came down out of control into the battlefield near British trenches. A night patrol recovered Schreiber body. The impact had buried Schäfer's body under the engine and attempts to recover it proved impossible.

Schreiber had been nominated for the award of the *Pour le Mérit*. In a letter to Schreiber's mother General von Below wrote on 5 July 1918, that the

Exceptionally vigorous spirit, the constant pushing forward, characteristics of the fallen fighter which he displayed time and again during numerous missions over the lines and which led to his great successes, were the determining factors in the Supreme Commander's recommending your son for the highest of all awards, the high Order Pour le Mérit. *It is an honour-laden duty for me to tell you that it pleased His Majesty the* Kaiser und König *on 8 June to award your son the high* Order Pour le Mérit.[30]

Schäfer had been nominated for the highest award for an enlisted man, the *Goldenes Militär-Verdienstkreuz* (Military Service Cross in Gold). It was an extra honour that the awards were made posthumously.

Fortunately, we have an eye-witness account of an action with a J.I. Charles Biddle went to France before the US entry into the war and flew for France flying Spad 7 fighters with *Spa* 73 before converting onto the Spad 13. After the US entry into the war he transferred to the USAS but remained with *Spa* 73 until he joined the Lafayette *Escadrille* that was at that point of time becoming the US 103rd Aero Squadron in February 1918. He was back to flying the single gun Spad 7 fighter. On 15 May 1918, Biddle was having a bad day. He had attacked an enemy two-seater and appeared to have put the gunner out of action and closed for the kill when his gun jammed. As his fuel was running low he turned the flight towards his lines where he saw "a peculiar two-seater" circling low over the trenches. At first Biddle thought it was a British aircraft but as it seemed to display remarkable impunity from the German lines he dove for a closer look.

The only way that I knew that my friend was really a Hun was by his crosses, for it was the first Boche machine of the kind that I had ever seen, and indeed I have never heard of anyone that I know, running into one like it. He had a rounded body like some French machines, the tail was square and the lower wing was much shorter than the upper, like many of the English two-seater observation planes. All the Hun two-seaters that I have ever seen or heard of before, have both upper and lower wings approximately the same length. In addition to this it was the slowest bus you ever saw and I think I could go two miles to his one. All this leads me to believe that it was a new type of German armored plane which they call "Junkers" and which I have read about in the aviation reports. They are built especially for this low infantry liaison work and are heavily armored about the fuselage to protect them from fire from the ground. In consequence of their great weight they cannot go very high and are extremely slow. ...

As soon as I was sure that the machine was really a Hun I dove down after him and made up my mind this time to get at good close range. I did and ended up fifty yards directly behind his tail and slightly below, but I made one bad mistake, a real beginner's trick which was the cause of all my troubles. I evidently was not quite far enough below him and I had not fired more than one or two shots when I got caught in the back draught from his propeller, which joggled my machine about so that anything approaching accurate shooting was an impossibility. I saw one bullet go three feet to one side of him and another several feet on the other side, so stopped shooting for a second to get in better position...I entirely failed, for the instant, to appreciate the very slow speed of the Hun. I was already close to him, and when I dove down and then pulled up to shoot, I found to my astonishment that I had overshot the mark and was almost directly under him. ...He started to swerve from side to side to get me out from under him so that the machine gunner could shoot, and I tried to stay under him, swerving as he did and at the same time slowing down my motor to the limit so as to try to let get ahead of me enough to allow me to start shooting again. The Boche and I were at this time about twenty yards apart and if he had only had a trap door in his bottom he might have brought me down by dropping a brick on my head. However, he did not need it. The Hun gave a twist which took me for an instant beyond the protection of the fuselage. It was only for a second or two, but that was sufficient for the observer, who proceeded to do the quickest and most accurate bit of shooting that I had yet run up against.

The gunner's first shot smashed through the Spad above the motor and hit Biddle's left leg just on top of the knee. His motor cut out and he dived down towards his own lines trying to get a much distance from the little height he had as possible. Coming down in the shell torn land the machine turned over onto its back. Biddle lost no time in scrambling out of the wreck and sheltering in a shell hole. While he was crawling away from his wrecked Spad Biddle noted that the Junkers was flying overhead and he thought it was shooting at his downed machine but he did not look to make sure.[31]

It appears that Biddle was the victim of Schreiber and the overflight after the combat could have been Schreiber taking his photographs.[32]

There were difficulties in repairing the J 4 as a letter from Seitz to Professor Junkers in May 1918 points out. Wagenführ had agreed that *Jco* should repair the J 4 machines "without knowledge of the manufacturing difficulties involved.... Also Junkers-Fokker-Werke have now realised the difficulties and have made a second suggestion that *Jfa* repair the wings and Junkers & Co the fuselage, fuselage parts and armoured tub. This also appears unacceptable." Unfortunately the letter gives no indication as to what were the difficulties involved in repairing the machines.[33]

In 1919 an improved version of the J 4 appeared that had the rear fuselage completely covered with corrugated metal rather than the fabric covering. This has been held to have had the military designation J.II but it has not been confirmed. It is thought that this was J.749/18 that had been modified in accordance with requirements of October 1918 where Junkers' proposal for a new armoured monoplane were rejected and a lightened J.I was proposed in order to utilise the parts that were already made rather than change the assembly line to a completely unproven machine. The armoured nacelle was redesigned to save weight and the wing span was increased.[34] Again the Allies were aware of the situation as a British Report of 21 October 1918, states that the Junkers Fokker Co had produced a new infantry machine called the J.II. "This is regarded more favourably by pilots than the J.I."[35]

Production of the J.I continued till at least January 1919. When production ceased Junkers had supplied 227 J.I biplanes out of orders for 283.[36] NASA, in its *Quest for Performance: The Evolution of Modern Aircraft,* considered the J.I was "a truly remarkable aircraft" that "has not received proper recognition in the literature of World War I aviation."[37]

Specifications for WWI aircraft are notoriously different and in the case of the Junkers products even more so as the reports produced by the British are not to their usual standards as they were obtained by measuring abandoned rather than captured aircraft. This is particularly true for the J 4, two damaged machines being the subject of the report.

Endnotes Chapter 5

1. The J-type designation came about due to the term "I" for *Infanterie* that in old German looks like a "J", and this was adopted such that it is always written as a J today.
2. For an appreciation of the development of German tactics and requirements for this type of work see "German Contact Patrols" by M. Meech, *Cross & Cockade International,* Vol.42 No.1, UK, 2011, P.26.
3. For the story of the German armoured aircraft see Herris, J. *Germany's Armoured Warplanes of WWI,* Aeronaut books, USA, 2012.
4. Doctor of Engineering (*Dr-Ing)* Otto Mader was born on 17 September 1880. A mechanical engineer from the University of Munich he joined Junkers at Aachen in 1909. With Hans Steudel he became technical director

of the Junkers factory during WWI. He was responsible for the initial study of Junkers aircraft designs. After the death of Otto Reuter he continued the development of Junkers aircraft with Ernst Zindel. He developed the aircraft that were built in Soviet Russia. He died in September 1944.

5. Otto Reuter was a design engineer who joined Junkers in November 1915. After the war he became the first technical managing director and chief engineer of Junkers Flugzeugwerke AG. He was responsible for the design of the F.13 and JG1 before his death in 1922 at only 35 years of age.
6. *Prof-Dr* Hans Steudel was born on 22 July 1883. In 1911 he joined Parseval in Berlin as a construction engineer before joining Junkers in 1913 by which time he was regarded as a leading engineer in the construction of airship engines. He was a technical director at Junkers during the war. After World War II he was forced to work in Russia for the Soviets. He died in 1963.
7. Franz Brandenburg attended the technical universities at Darmstadt and Braunschweig between 1902 to 1908. He became a secretary assistant to a Professor Kochy at the University of Aachen in 1910, being responsible for railways and mass transport questions. Joining Junkers in 1912 he was a design engineer and after WWI undertook financial duties for *Junkers Motorenwerke GmbH.* He died in December 1970 at the age of 88 years.
8. According to the US Attaché's Report of 26 April 1921, he had received three advertising catalogues of the Junkers Company. "In regard to the J I type built for the Army during the war, Professor Junkers stated that the Army requirements specified a biplane type and that he had to put on the lower wing to satisfy the Army, but he did so with great reluctance, and only after objecting to such construction as forcibly as he was in a position to." Copy Attaché's Report No. 58 in P.M. Grosz Collection Box 483.
9. The *ZAK* Report states that three were ordered. Prototypes were often ordered and then purchased at a later date having a higher serial number than the production models.
10. Reinhard Zankl suggests that the contract was cancelled as no serials were issued.
11. *Junkers Flugzeugwerke AG,* also abbreviated as *Jco,* came into existence on 24 April 1919.
12. According to Dan-San Abbott, Staaken and Junkers had a monthly production capacity of 30 aircraft per month and this related to the J 4. E-mail from Dan-San Abbott.
13. Fokker, A.H.G. & Gould, B. *Flying Dutchman – The life of Anthony Fokker,* George Routledge & Sons, Ltd. UK, 1931. PP.155-156. In his *Fokker - The Man and the Aircraft* (Harleyford, UK, 1961), Henri Hegener wrote that the enforced marriage of Fokker and Junkers broke up through incompatibility. Fokker was 27, full of boyish enthusiasm, with little scientific background, while Junker was a 57 year old distinguished professor and researcher who showed little understanding of the practical requirements of simplicity for wartime construction. Although Fokker acknowledged Junkers as the "father of the thick wing", Hegener cannot let his hero Fokker, be cheated out of the accolades for adopting the thick wing. He disputes Junkers' claim, pointing to the Frenchman Léon Levavasseur who built a monoplane in 1911 with a cantilever thick wing. This machine never flew as the engines of the time could not develop enough power. (Hegener P.33.) Hegener cites Rhinold Platz as one who should know. Platz's claims are still subject of controversy. See Sunderland, G. "Designing Fokker Aeroplanes," and Owers, C.A., "The Fokker-Weyl Controversy"' both in *Over the Front,* Vol.30 No.3, 2015.
14. Jane's recorded that the "Junkers-Fokkers Works, Ltd," was "One of the most interesting new formations of 1917" founded by "the co-operation of the well-known concerns of the aeroplane manufacturer, Fokker, and the Diesel engine manufacturer, Junkers. The subtitle of "metal aeroplane building indicates possibly the field of activity." (*Jane's Fighting Aircraft of World War* I, Studio Editions, UK, 1990. P.167.) As described above there was no co-operation between Fokker and Junkers, It was a forced marriage and soon ended in divorce.
15. Byers, R.W.E. *Power and Initiative in Twentieth Century Germany, The Case of Hugo Junkers,* Dissertation for Doctor of Philosophy, University of Georgia, 2002. P.38.
16. Summary of Technical Air Intelligence, Wednesday, 6th November 1918. Compiled by A.1.4. (Directorate of Air Intelligence). Copy in NARA RG45 Entry 520 Box 142.
17. According to Peter M. Grosz there is no proof that any Junkers J.I was shot down behind Allied lines. Grosz, P.M. *Junkers J.I,* Albatros Productions, Windsock Datafile No.39, UK, 1993.

18.The upper port and starboard wings differ in internal structure and may have been made by different factories. This conclusion comes from the use of fabricated 'U' channel bracing between spars on the starboard wing and round tubing on the port. Additionally, the port wing has numerous internal markings specifically on all of the spar tubes whereas the starboard wing has none of these markings. Both wingtips have airframe ID stencils showing the same number, 586. (Notes by John Weatherseed). As far as is known only *Jfa* built the production J 4.

19.Summary of Technical Air Intelligence, Wednesday, 6th November 1918. Compiled by A.1.4. (Directorate of Air Intelligence). Copy in NARA RG45 Entry 520 Box 142.

20."Junkers postscript " "Post Box", *Airfix Magazine,* August 1983. P.548.

21.Quoted in Penrose, *British Aviation, The Great War and Armistice.* Putnam, UK, 1969.

22."Enemy Aviation for the Month of January, 1918." Ministry of Munitions. TNA AIR1/2094/207/12/8. A meeting of the Chief of Air Staff (CAS) of 22 January 1918, was asked if any armoured German machine had been encountered and if so have they been attacked by scouts. The answer was that no German armoured machines had yet appeared, presumably meaning that the British had not encountered any. TNA 1/6A/4/37.

23."Enemy Aviation for the Month of February, 1918." Ministry of Munitions. TNA AIR1/2094/207/12/8.

24."Enemy Aviation for the Month of April 1918 – Translation of Monthly Report, issued by the Technical Section of Military Aeronautics under the Minister of War, French Army." P.3. TNA MUN4/5010.

25."Enemy Aviation for the Month of May, 1918." Ministry of Munitions. TNA MUN 4/5010.

26.Grey, P & Thetford, O. *German Aircraft of the First World War,* Putnam, UK, 1962, P.156.

27.*Cross & Cockade* (USA), Vol.3 No.3, P.271.

28.From the Gauting Archive. We are indebted to German historian and author Bruno Schmäling who provided the reports, and to Roland Jahne and Klas Bask who provided the translations.

29.Hptm Oskar Minarski was the CO of FA (250(A) from about May 1917 to 15 December 1917.

30.Quoted in Rabe, H-R. "A Forgotten Pour le Mérit Flyer – Leutenant Wilhelm Paul Schreiber." *Cross & Cockade* USA, Vol.15, No.2, Summer 1974. P.185.

31.Biddle, C.J. *The Way of the Eagle,* Scribners' sons, NY, 1919.

32.Thanks to Greg Van Wyngarden for referring me to the research of Dennis Connell, Bill Bailey and Jon Gutman that ties the action of Schreiber and Biddle together. Although it can never be fully confirmed it appears that, given the location of the 103rd Aero Squadron and losses that day, it is most probable that this is so. See Connell's letter in the "Wind Soc" column in *Cross & Cockade* (USA), Vol.16 No.2, 1975, P.188; and Bailey's history of the 103rd Aero Squadron in *Cross & Cockade* (USA), Vol.19 No.4, 1978, P.302.

33.Seitz to Professor Junkers, 17.06.1918. In the following paragraph Seitz recorded that a meeting would be held with *Idflieg* with representatives of Germania, Rinne, *Jfa* and *Jco.* Seitz considered it important that Germania and Rinne were "provided with drawings at the right moment so that every appearance of a delay to the detriment of the Army is diverted." Whether this was to do with the J 4 repairs is not known but it appears that these firms were to be brought into manufacturing Junkers parts at least. There are also hints within this correspondence of strained relations between the military and the Junkers concerns. Seitz then recorded that "Fokker is said to have unchallenged been awarded an order for 200 parasols."

34.P.M. Grosz wrote that the *Idflieg* Status Report for September 1918, delivered 8 November 1918, has a J.II listed, but gives no details. It also lists an AEG J.III. The AEG J.III had very similar dimensions as to its forerunner but had a D.IVa engine and new armour. P.M. Grosz to J.M. Bruce, 20.11.1969. RAF Museum J.M. Bruce Collection Box 70.

35.TNA AIR1/474/15/312/184.

36.P. Leaman in his article "The Fokker-Junkers Seaplanes" notes that "about 184 were delivered." *Cross & Cockade International,* Vol.46 No.2 2015, P.104.

37.*Quest for Performance: The Evolution of Modern Aircraft.* Chapter 2. Design Exploration 1914–18. Army Cooperation Aircraft. NASA's History Office.

Junkers J 1

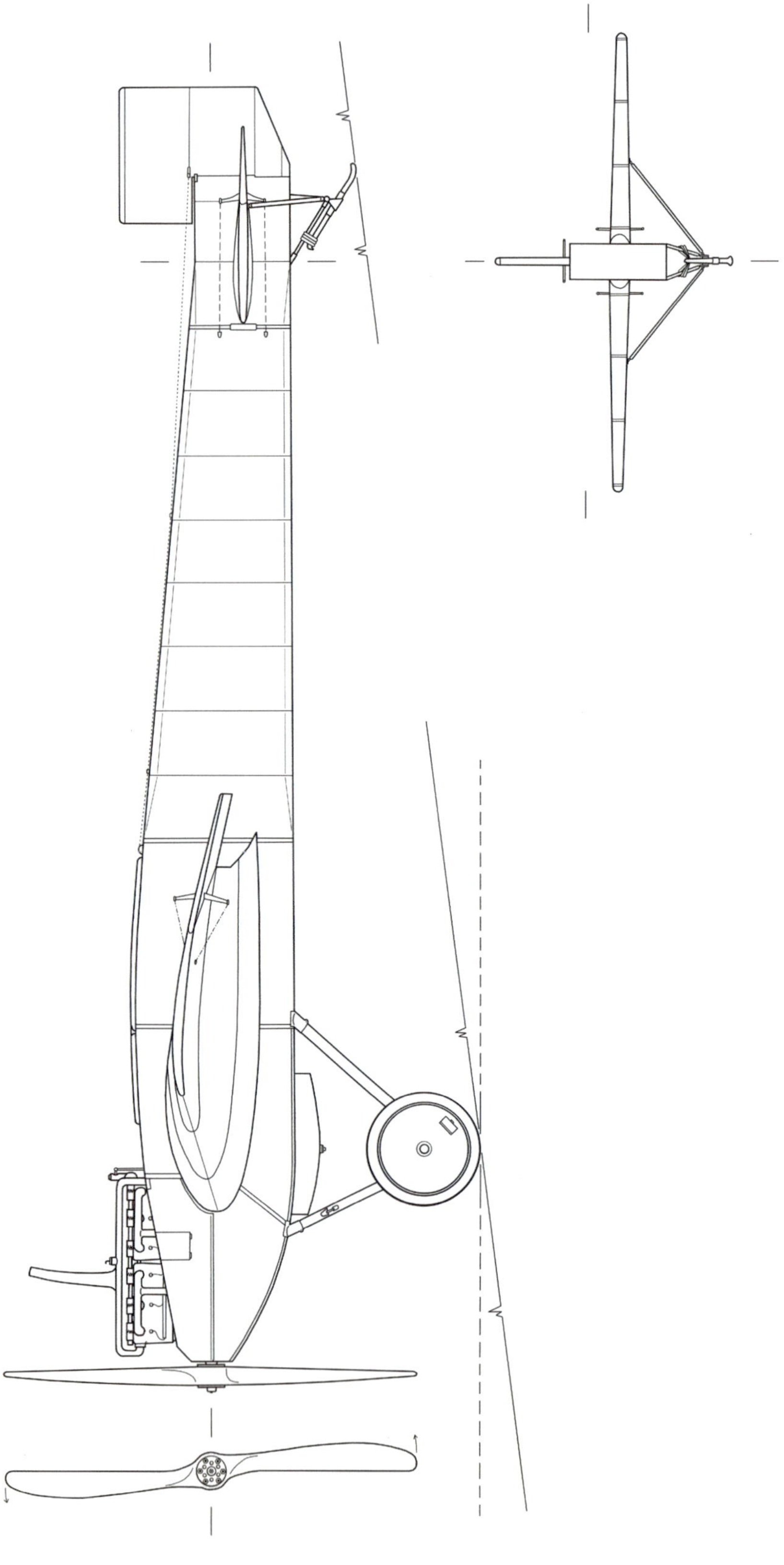

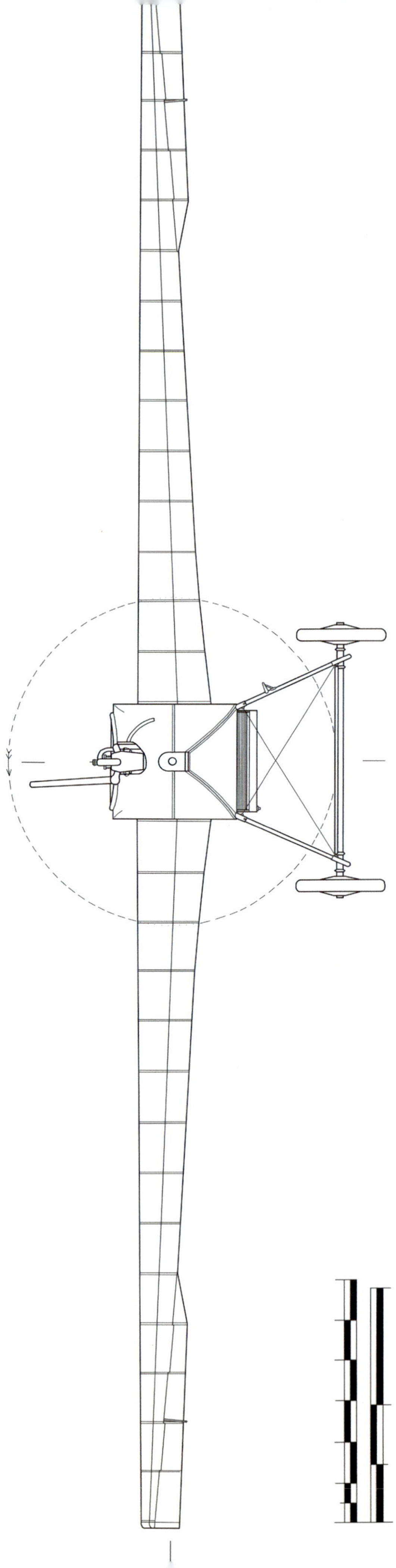

Junkers J 1

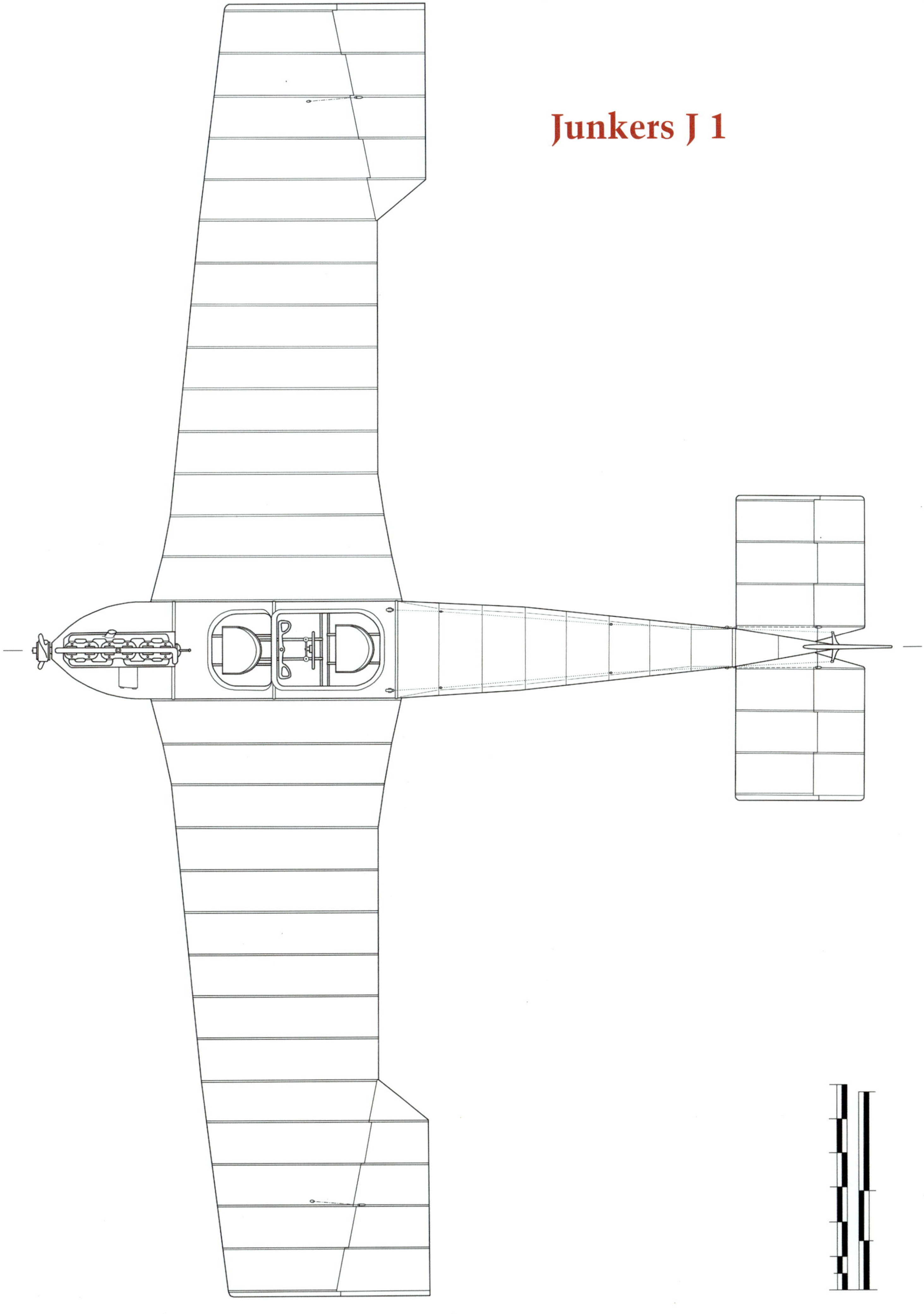

Junkers J 1

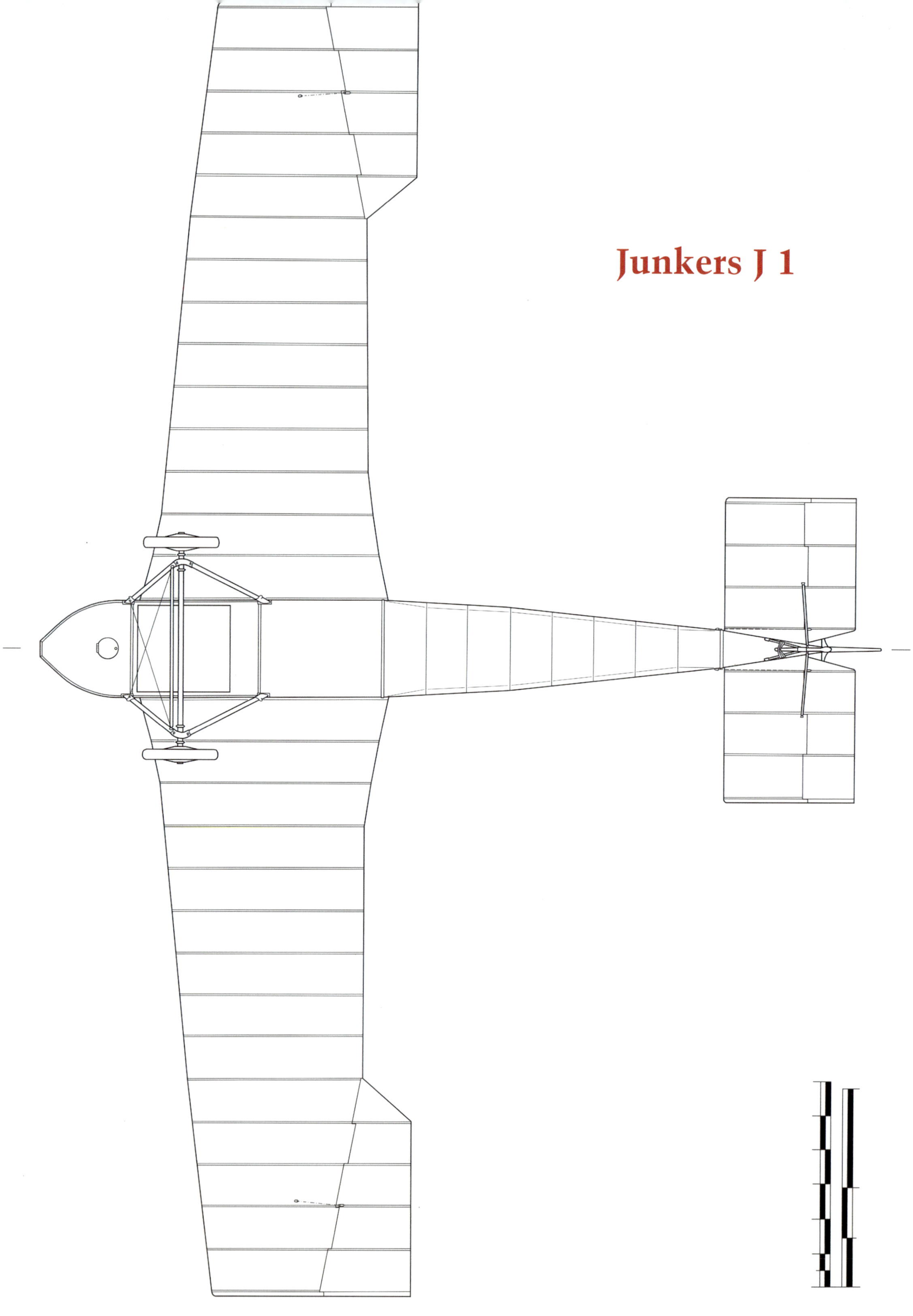

Junkers J 2 (E.II)

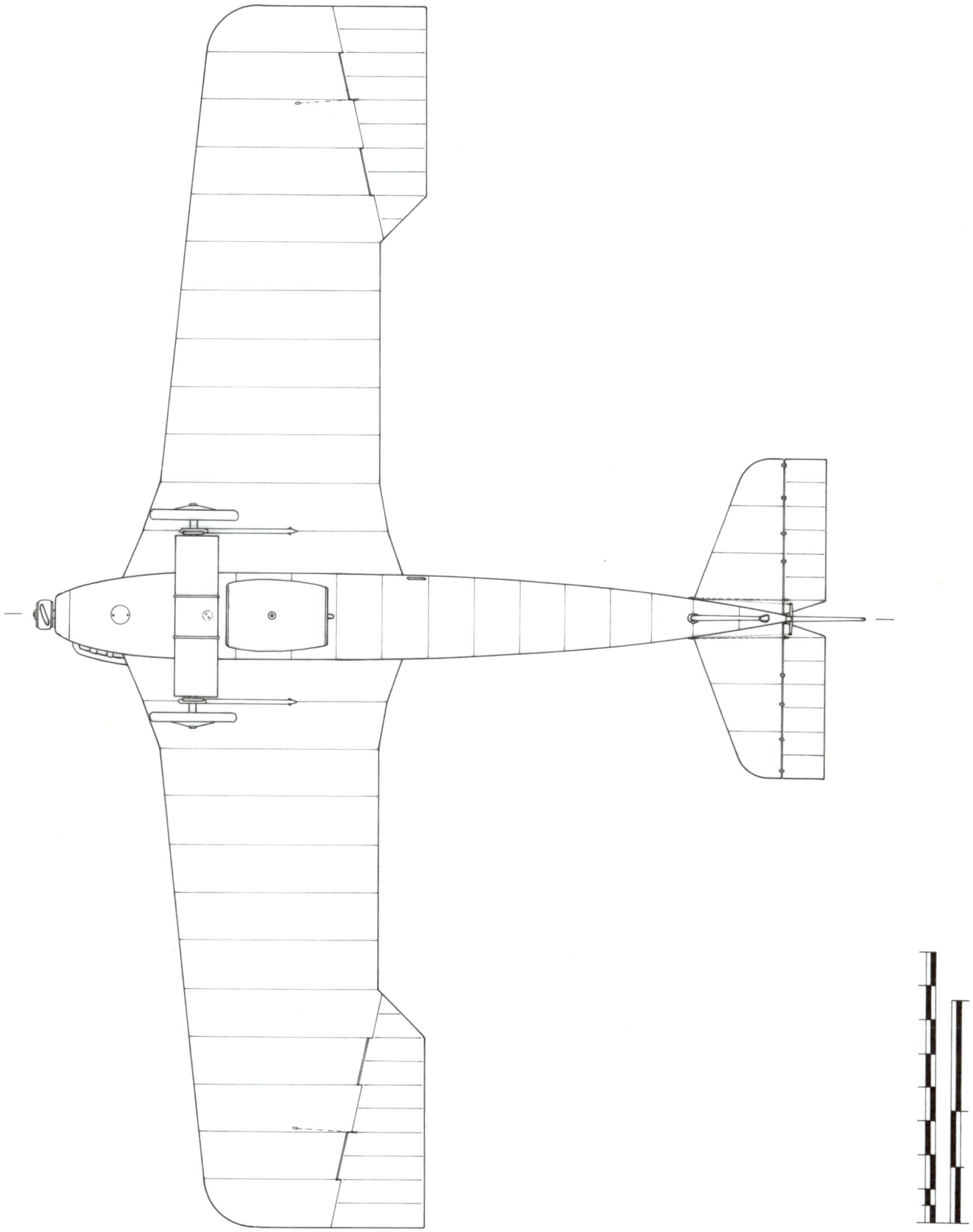

Junkers J 2 (E.II)

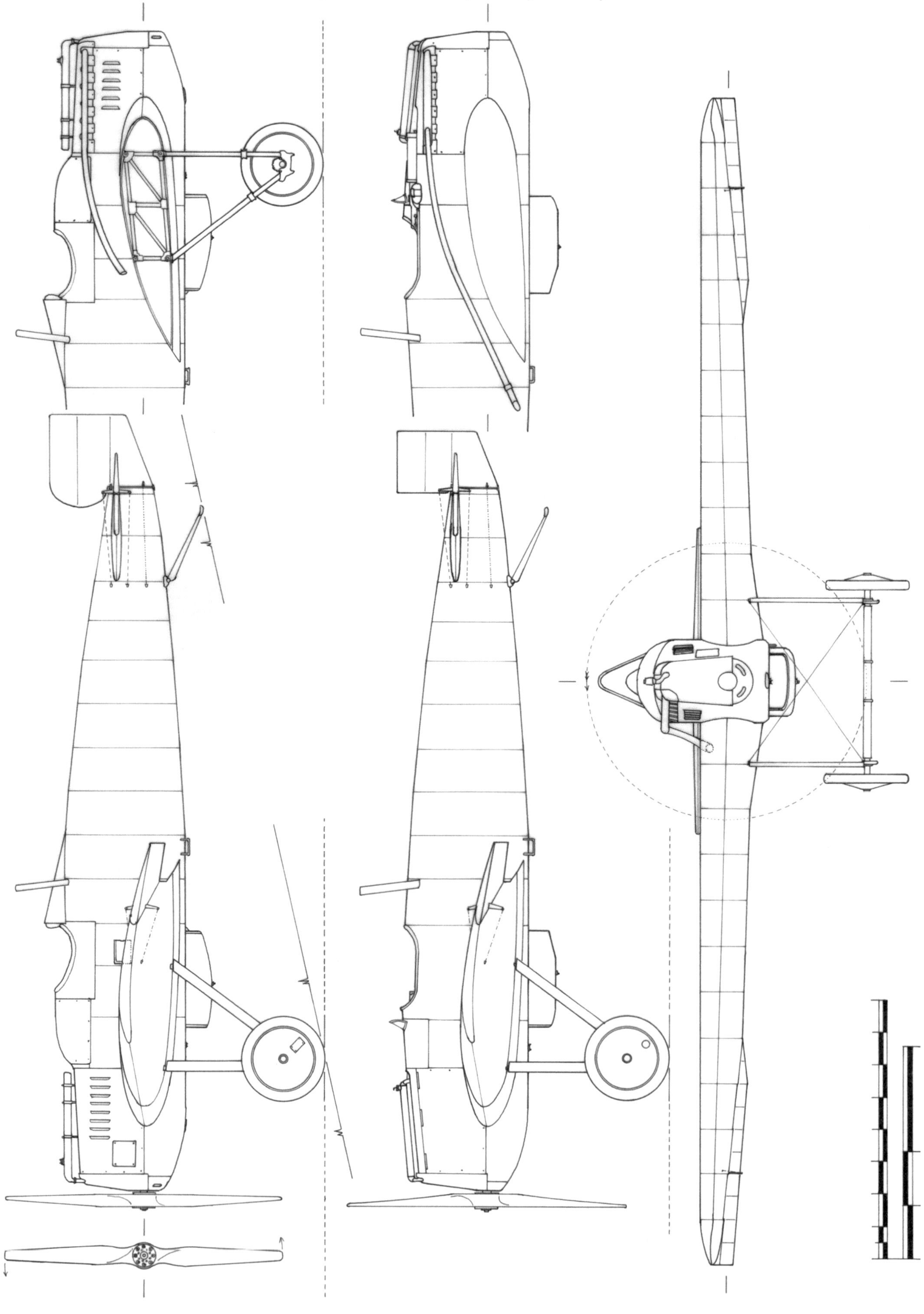

Junkers J 2 (E.II)

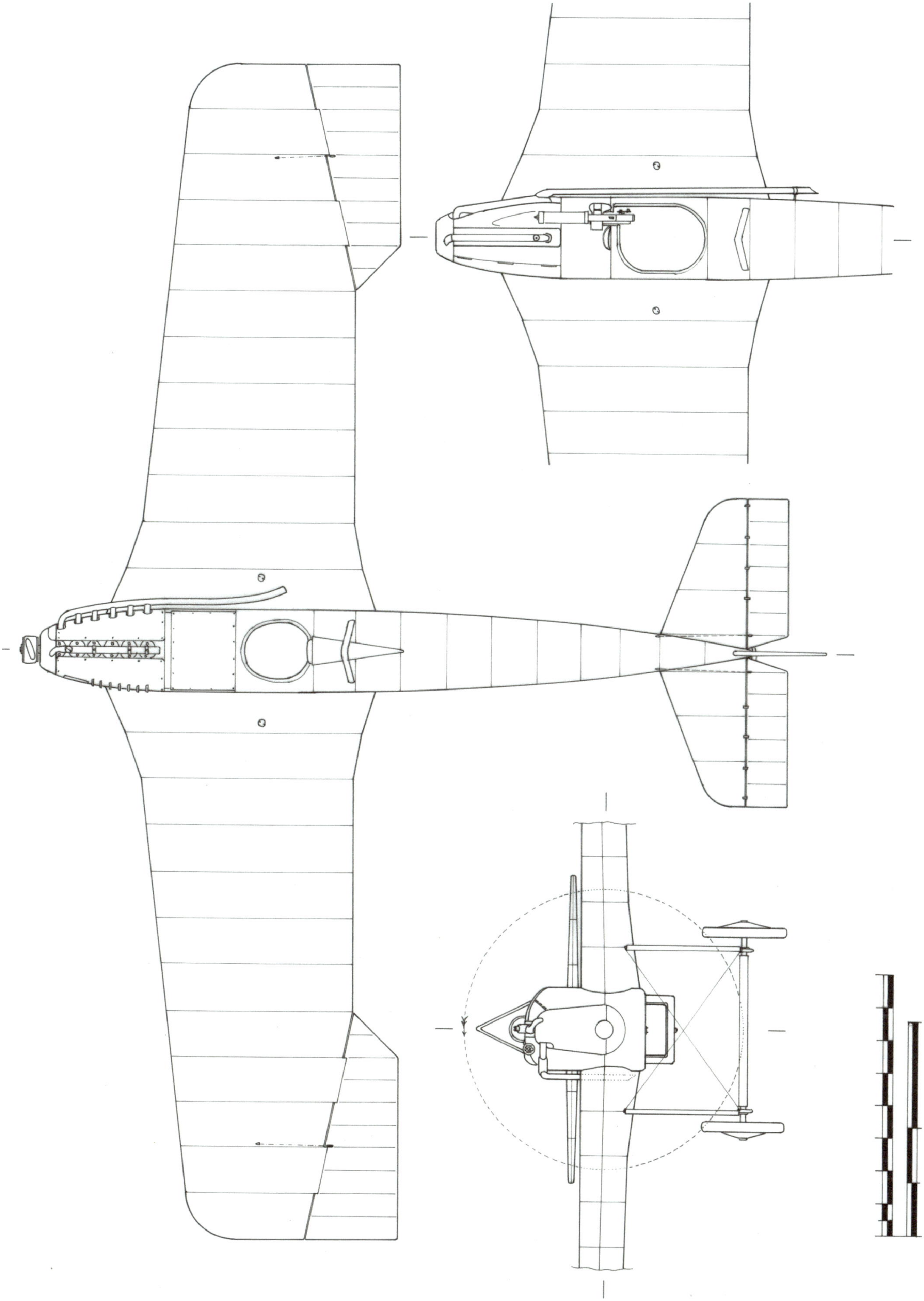

Junkers J 4 (J.I)

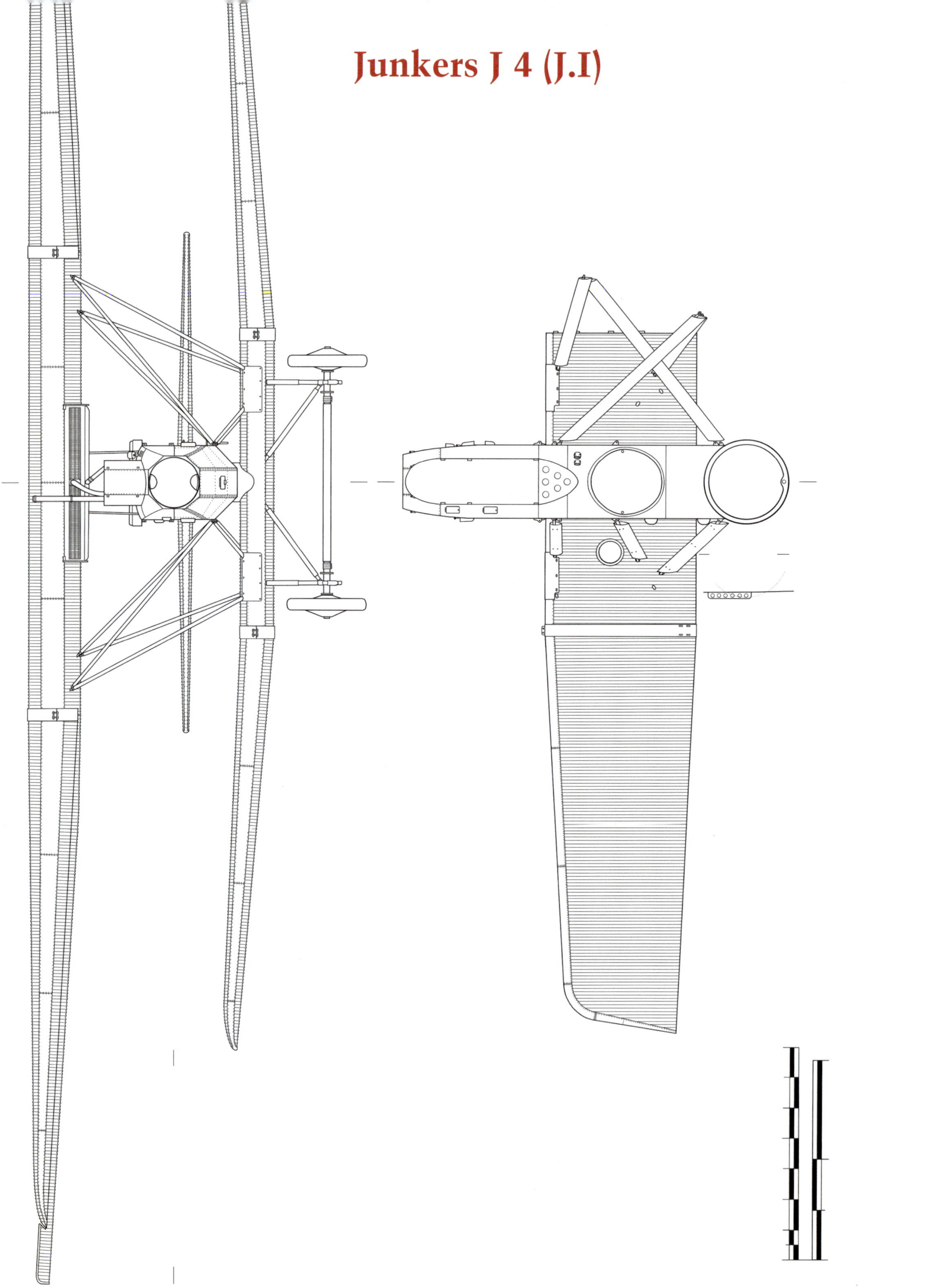

Junkers J 4 (J.I)

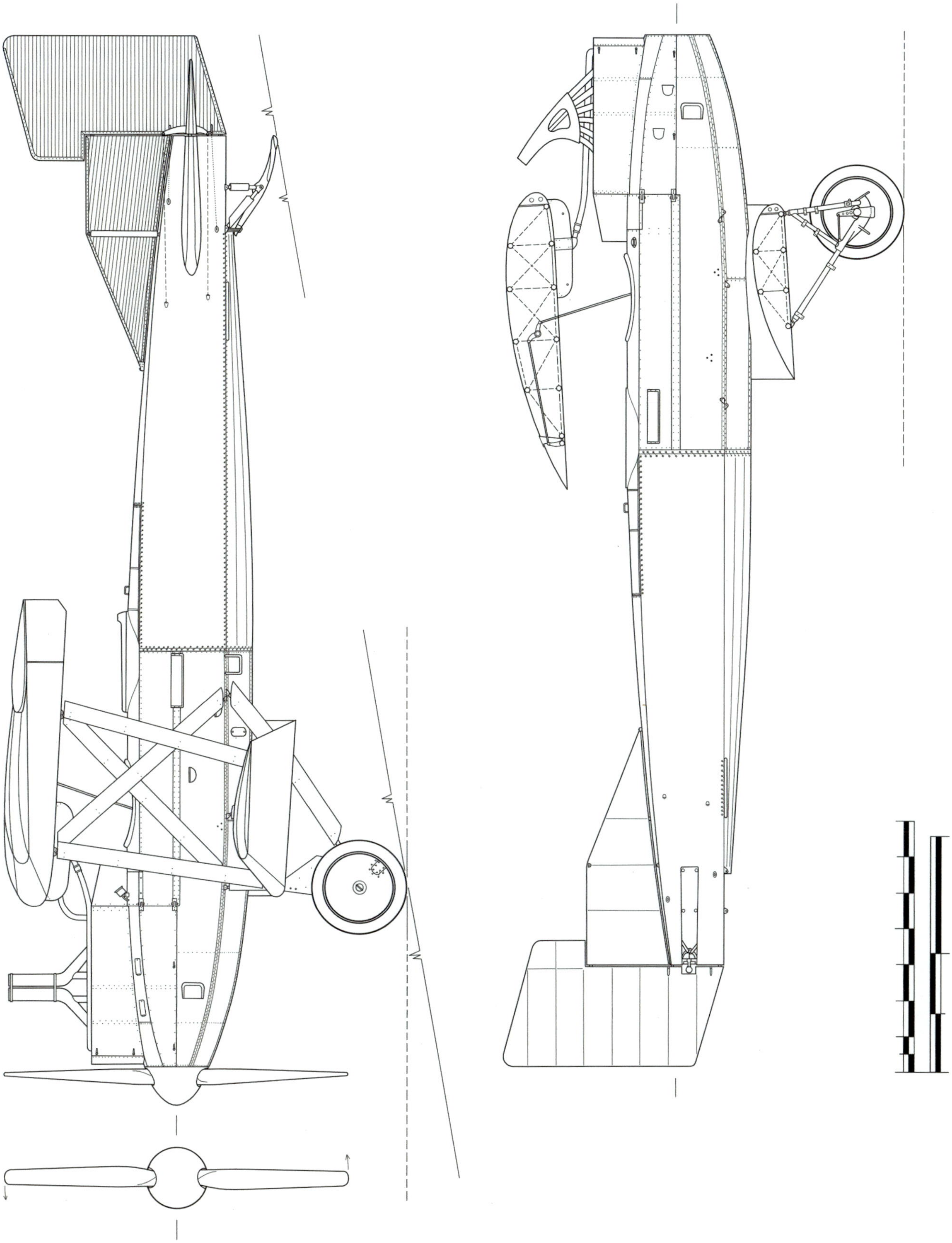

Junkers J 4 (J.I)

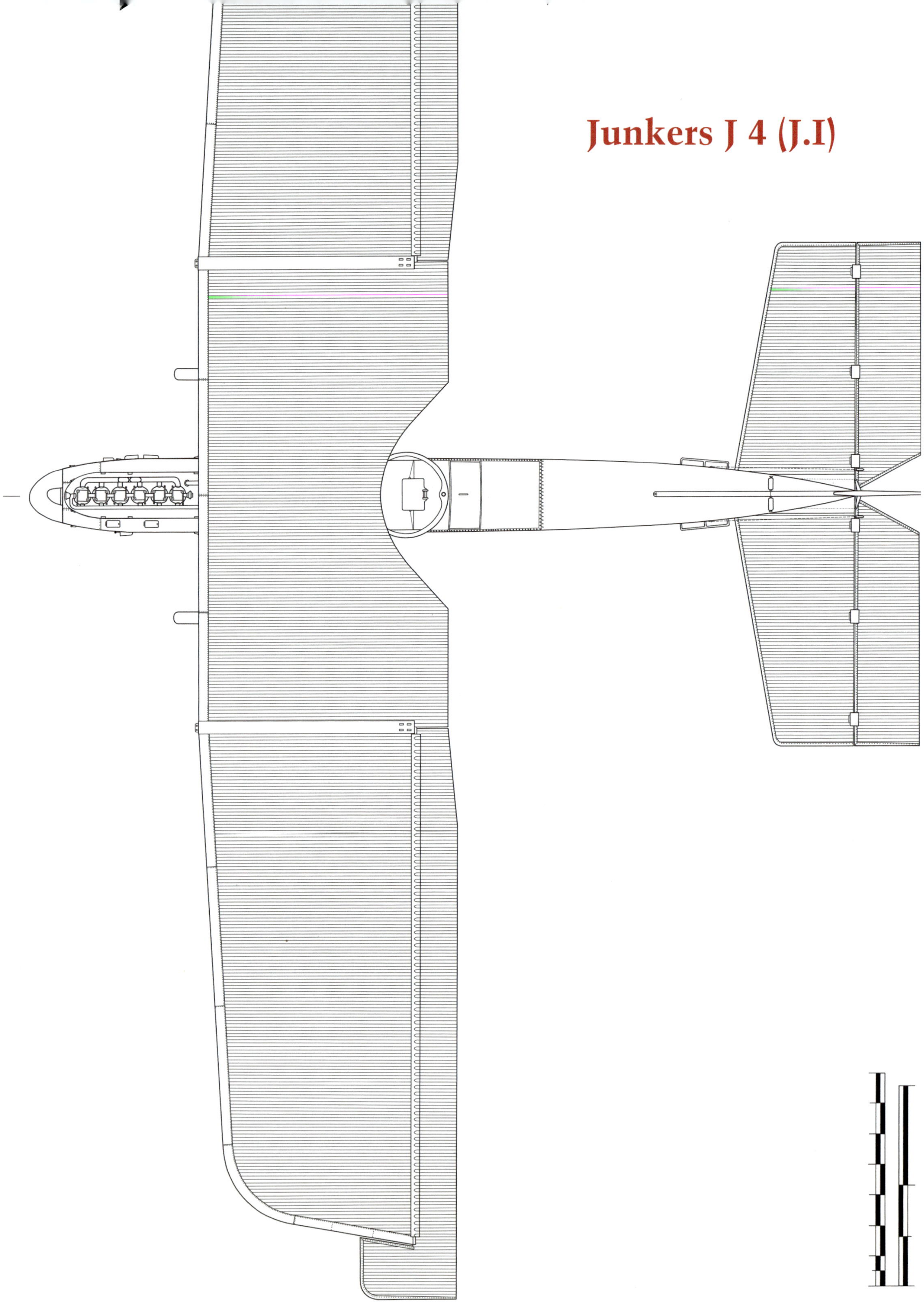

Junkers J 4 (J.I)

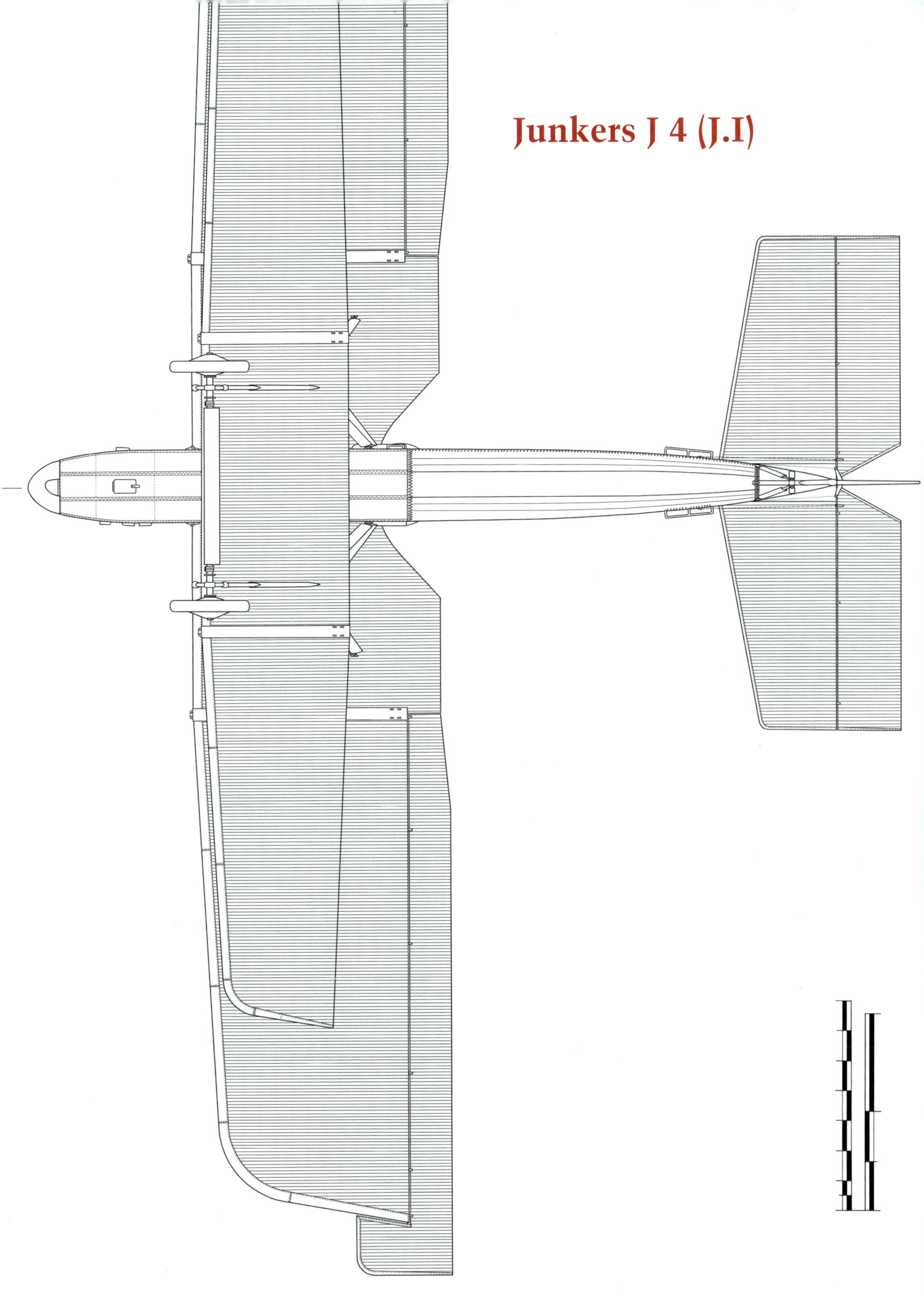

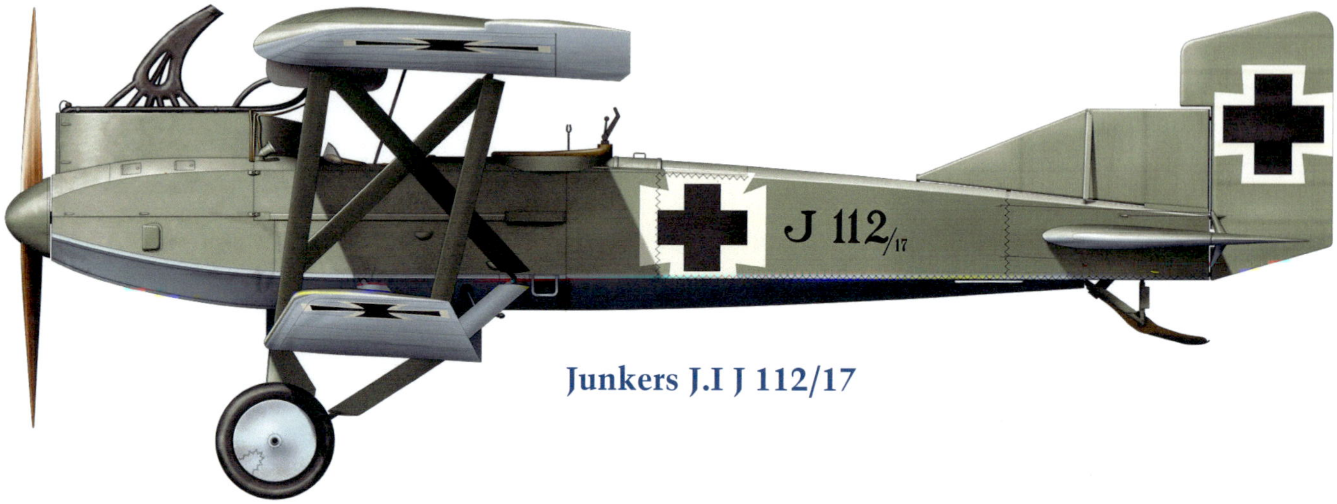

Junkers J.I J 112/17

Junkers J.I J 138/17

Junkers J.I 802/17

Made in the USA
Las Vegas, NV
06 November 2026

57642758R00126